Kinanthropometry X

The 10th International Conference of the International Society for the Advancement of Kinanthropometry (ISAK) was held in conjunction with the 13th Commonwealth International Sport Conference (CISC2006), Melbourne, Australia, 9–12 March, 2006, immediately prior to the XVIII Commonwealth Games. This volume contains a selection of papers presented to the Conference, covering the following topic areas:

- body composition and bone density
- athlete morphology, sexual dimorphism, somatotype and performance prediction
- 3-dimensional analysis and body sizing
- virtual anthropometry
- body image
- anthropometric pedagogy.

These papers represent the current state of research and knowledge in kinanthropometry, and will be of particular interest to students and researchers in sport and exercise science, kinanthropometry, physical education and human sciences.

Mike Marfell-Jones is Professor and Chair of Health Sciences at the Universal College of Learning, Palmerston North, New Zealand and President of the International Society for the Advancement of Kinanthropometry.

Tim Olds is a Professor in the School of Health Sciences and Director of the Centre for Applied Anthropometry at the University of South Australia.

Kinanthropometry X

Proceedings of the 10th International Society for the Advancement of Kinanthropometry Conference, held in conjunction with the 13th Commonwealth International Sport Conference

Edited by

Mike Marfell-Jones and Tim Olds

Routledge
Taylor & Francis Group
LONDON AND NEW YORK

First published 2008 by Routledge
2 Park Square, Milton Park, Abingdon, Oxon OX14 4RN

Simultaneously published in the USA and Canada
by Routledge
270 Madison Ave, New York, NY 10016

*Routledge is an imprint of the Taylor & Francis Group,
an informa business*

Note: This book was prepared from camera-ready copy supplied
by the editors

Printed and bound in Great Britain by MPG Books Ltd, Bodmin

British Library Cataloguing in Publication Data
A catalogue record for this book is available from the British Library

Library of Congress Cataloging in Publication Data
International Society for Advancement of Kinanthropometry. International
Conference (10th : 2006 : Melbourne, Australia)

Kinanthropometry X : proceedings of the 10th International Society for
Advancement of Kinanthropometry Conference, held in conjunction with
the 13th Commonwealth International Sport Conference / [edited by]
Mike Marfell-Jones & Tim Olds.

p.; cm.

Kinanthropometry 10

Kinanthropometry ten

Includes bibliographical references and index.

ISBN 978-0-415-43470-6 (hardcover)

1. Sports--Physiological aspects--Congresses. 2. Kinesiology--Congresses.
3. Anthropometry--Congresses. 4. Somatotypes--Congresses.
I. Marfell-Jones, Mike. II. Olds, Tim. III. Commonwealth International
Sport Conference (13th : 2006 : Melbourne, Australia) IV. Title.
V. Title: Kinanthropometry 10. VI. Title: Kinanthropometry ten.

[DNLM: 1. Anthropometry--Congresses. 2. Body Composition--
Congresses. 3. Body Constitution--Congresses. 4. Body Weights and
Measures--Congresses. 5. Sports--Congresses. GN 51 I61 2007]

RC1235.I525 2007

613.7--dc22

2007005572

ISBN 978–0–415–43470–6 hbk

Contents

Preface

The International Society for the Advancement of Kinanthropometry (ISAK) is an international group that has made its role the stewardship and advancement of kinanthropometry – the study of human size, shape, proportion, composition, maturation and gross function. Kinanthropometry, a discipline of only some forty years standing, is named from the Greek root words kinein (to move), anthropos (man) and metrein (to measure). Its purpose is to facilitate a better understanding of growth, exercise, performance and nutrition. Although ISAK is probably best known for its international anthropometry accreditation scheme, it also plays a major role in increasing international dialogue between kinanthropometrists and facilitating the presentation and dissemination of the results of their scientific endeavours.

To that end, for the past twenty years, ISAK has made a practice of holding its Biennial General Meetings in association with major sport science conferences prior to either the Olympic or Commonwealth Games. Within these conferences, ISAK has been instrumental in the organisation of a kinanthropometry stream as a platform for the presentation of quality papers on the research findings of its members and international colleagues. In the past decade, that has occurred in Adelaide in 1998 (prior to the XVI Commonwealth Games in Kuala Lumpur); in Brisbane in 2000 (prior to the Sydney Olympics); in Manchester in 2002 (prior to the XVII Commonwealth Games) and in Thessaloniki in 2004 (prior to the Athens Olympics. Subsequent to these conferences, ISAK appointed editors to produce a set of proceedings so that these papers could be accessed by a far wider audience than those able to attend the conferences themselves.

In March 2006, this tradition continued in Melbourne, Australia, in conjunction with the 13th Commonwealth International Sport Conference, immediately prior to the XVIII Commonwealth Games. These proceedings, which represent a refereed selection of the best kinanthropometry papers presented in Melbourne are the tenth in the series. ISAK thanks the editors most sincerely for their efforts in compiling this volume and congratulates the authors on their work and its inclusion in this publication.

Mike Marfell-Jones
President
International Society for the Advancement of Kinanthropometry

Introduction

M. Marfell-Jones[1] and T. Olds[2]
[1]UCOL, Private Bag 11 022, Palmerston North, New Zealand
[2]Centre for Applied Anthropometry, University of South Australia,
City East Campus, GPO Box 2471, Adelaide SA 5001, Australia

The contents of this volume represent the Proceedings of Kinanthropometry X, the 10th International Conference on the subject area of kinanthropometry. The conference was incorporated within the 13th Commonwealth International Sport Conference, in Melbourne, Australia from 9 to 12 March 2006, the week preceding the XVIII Commonwealth Games. The kinanthropometry strand of the conference was organised by the International Society for the Advancement of Kinanthropometry (ISAK) in conjunction with the conference hosts, Sports Medicine Australia, under the guidance of the Conference Organising Committee Chair, Frank Pyke.

All presenters whose papers had a kinanthropometric focus were invited to submit full manuscripts for consideration for selection. Not all communications were worked up into manuscripts for the Proceedings and not all of those that were written up succeeded in satisfying peer reviewers. The selected chapters are from across the kinanthropometric spectrum.

The monograph opens with chapters on anthropometric pedagogy and maturation. These are followed by five chapters presenting the results of different investigations using whole-body scanners to evaluate 3D anthropometry. A second group of five follows which reports on body composition in elite athletes with football, rowing, water polo and rugby featuring. Chapter 13, a technical paper on subcutaneous adipose tissue measurement, is followed by a rarity in anthropometric circles – an epidemiological chapter. Chapter 15 introduces the concept of a unique anthropometric language. The penultimate chapter examines the increasingly-important area of body dissatisfaction and the final chapter looks at proportionality in elite swimmers. The international contributions both provide readers with an outline of the current state of knowledge in kinanthropometry around the globe and reflect a broad interest in the subject area world wide.

The editors are grateful to the contributing authors for their positive involvement in the compilation of this book, particularly their responsiveness to the review process, and to the reviewers. We also acknowledge the invaluable technical assistance of Ms Suzanne Little, Faculty of Health Sciences at the Universal College of Learning, for compiling the camera-ready copy of the text.

Mike Marfell-Jones
Tim Olds

CHAPTER ONE

Pedagogic approaches to teaching anthropometry

A.D. Stewart

School of Health Sciences, The Robert Gordon University, Aberdeen, UK

1 INTRODUCTION

Anthropometry is the science of making surface measurements on the human body based on anatomical landmarks. It is recognised as a discipline in its own right in describing the human phenotype, and interfaces with a range of others including biomechanics, physiology and nutrition to have a role in assessing health and sports performance. Despite their apparent simplicity, acquiring the skills for making skinfold, girth or skeletal breadth measurements can be problematic, because a large number of simultaneous tasks need to be performed in a movement which lasts only a few seconds. Evidence of this complexity is readily apparent from the ten hours of practical tuition the International Society for the Advancement of Kinanthropometry requires of its instructors to induct professionals into the 17 measurements of the restricted profile (Marfell-Jones *et al.*, 2006). However, other organisations or individuals may fail to do justice to the intricacy of measurement, either by failing to describe methods in sufficient detail for techniques to become truly standardised (WHO, 1989), or by suggesting that attempts to do so are of 'mistaken exactitude' in view of errors in predicting body composition from anthropometric data (Durnin, 1997). Since this time, with the advent of a global protocol (ISAK, 2001), the first concern has been addressed, and an increased emphasis on retaining raw data, rather than convert it into tissue masses or percentages (Marfell-Jones, 2001) has largely addressed the latter.

Of concern to all practitioners in science is the reproducibility of methods and data, so a procedure can be replicated by different personnel in a different setting. With anthropometry, those who quantify the error of repeated measurements become aware of the pivotal role of standardisation in enhancing intra and inter-tester reliability. Relative to skinfold measurements made by an experienced tester, 30 minutes' tuition has been shown to produce a substantial improvement in inexperienced testers measurements when compared with those of a group not receiving instruction (Kerr *et al.*, 1994). Fitness instructors who routinely make skinfold measurements have appeared to show reasonable intra-observer reliability, but poor inter-tester reliability (Webster, 2002) highlighting the necessity for a standardised approach to measuring sites, and development of the required skill of the anthropometrist (Wang *et al.*, 2000). Such standardisation upholds 'best practice' in terms of demonstrable skill to be imparted during teaching.

Experienced anthropometrists including instructors may be between five and ten times more precise, accurate and quick at measuring a profile as novices in practice. Not only are instructors' movements 'automatic', they appear fluent, and one measurement 'flows' into the next, with minimal time gap. By contrast, a novice appears tentative, jerky, and separate movements are separated by pauses. In terms of measuring, there is a stark contrast between novice and instructor. In terms of teaching, the instructor is only simulating measuring, and will demonstrate 'end-form' technique, encourage practice, and correct faults. In practice, this may be more difficult than it sounds, as learners identify with different aspects of the skill, or interpret the instruction in different ways. Because students of anthropometry are likely to have different learning styles, an instructor may select a variety of methods to deliver a teaching experience to a group. The aim of this summary is to provide basic theoretical background for skill acquisition in anthropometry and illustrate some methods which may be appropriate for this process.

Skills are tasks which seek to achieve specific goals. They can be *closed* – predictable or stationary and not variable with other external input, or *open* – where changing or dynamic environmental conditions influence the requirements. In reality, there is a continuum between open and closed skills, and making anthropometric measurements is represented near the 'closed skill' end. However, the variability in plasticity of body tissues, measuring equipment and the physical measuring environment all contribute to variability.

Various theoretical models of acquisition of a practical skill have been proposed, for example the model of Fitts and Posner (1967), which details the awareness phase, practice phase and acquired skill phase of a physical task. The *awareness phase* involves participants understanding the basic principles and some of the complexity of the task. The *practice phase* involves developing skill at coordinating the different components. Both these phases are addressed by the techniques mentioned below. The *acquired skill phase* involves less attention being given to the task, to enable other tasks to be accomplished. In terms of a practical skill, this may involve experienced technicians working on a large measuring project or teachers whose skills are sufficiently subconscious to enable their attention to focus on the learner and the lesson. During effective learning, feedback is fundamentally important, to motivate, to reinforce positive and discourage negative actions. This can be *intrinsic* – the sensory information produced as a consequence of performing the movement, or *extrinsic* – additional information provided by peers, an instructor, video or the measurement value itself. The learner can sense how a measurement should feel using intrinsic feedback, while the instructor can control the extrinsic feedback, taking care that the right amount of information is provided in the right way on an individual basis.

How can this theoretical framework be applied in the teaching and learning of anthropometry? Experience from teaching other practical skills can reveal a variety of effective approaches which guide individuals through a journey of practical skill acquisition. A large array of techniques can be applied when introducing beginners to skill learning. Such methods are in use in ski teaching (BASI, 1997), and in a methodological context, there is no compelling reason to suggest that such methods cannot be applied in the development of anthropometric measuring skill.

A sequence of key stages can be depicted in anthropometric measurement, using the example of making a skinfold measurement. Such a sequence is depicted in Figure 1.1 A–F.

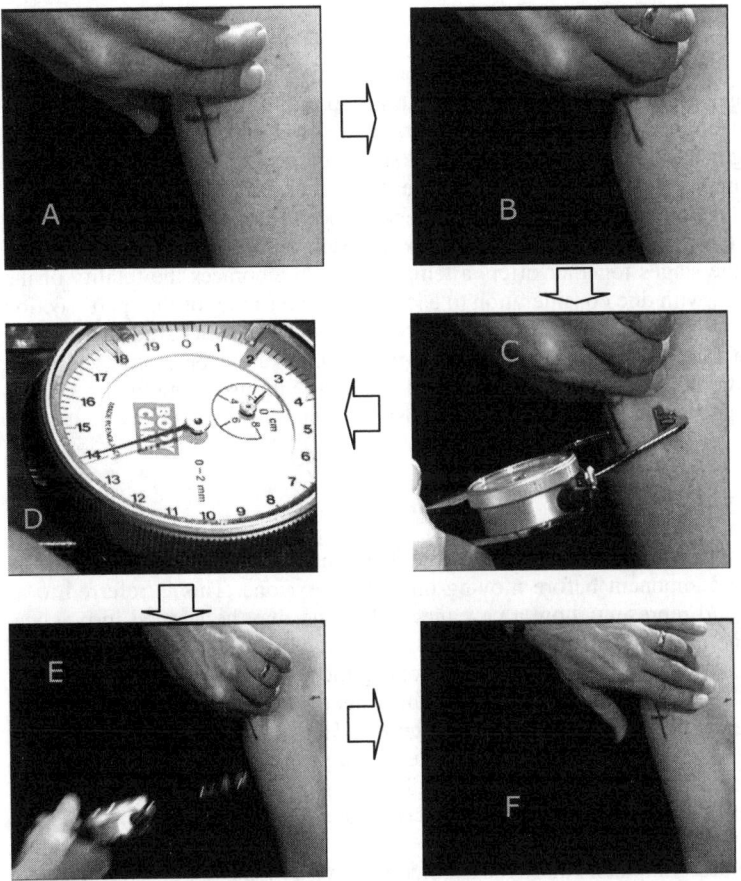

Figure 1.1 Stages of a skinfold.

A: Approach the skin with fingers and thumb at 90° to the skin surface
B: Align the finger border with the landmark, grasping enough, but no more than enough to raise a skinfold with parallel sides
C: Apply the calipers 1cm away from the edge of the finger, at approximately mid-fingernail depth
D: Release the caliper spring tension and take the reading after two seconds
E: Depress the trigger and remove the calipers, maintaining the skinfold grasp
F: Release the skinfold

These stages represent a considerable learning challenge to any novice, because so many factors have to be accomplished appropriately and in the correct sequence within only a few seconds. The difficulty of such a challenge is readily exacerbated by failing to clear sufficient space, or to use the anthropometric box to align the relative size of measurer and subject appropriately. Once these have been done, the tuition of the main part of the task still needs to be carefully considered, and can follow several alternative learning pathways.

2 WHOLE-PART-WHOLE

This involves first demonstrating the entire action, and then breaking it down into a series of stages. The sequence is gradually built up into the whole again. This may be a useful approach, because the fluency of an experienced demonstrator can readily disguise the complexity of the skill sequence. Presenting a clear picture of the requirements to students may assist their conceptual understanding, but many will fail to take on board all the steps in the sequence. Therefore, highlighting each stage becomes important, emphasising why each stage is included. Finally, 'stitching' the stages together offers a template which embraces the totality of the measurement, with due consideration to each stage. The timing of the 'part' section can be as much as 30 seconds, and this time can be reduced as students become more fluent, and eventually become 'real-time'. Alternating between 'real-time' speed and the 'slow motion' to identify all the parts is a useful method for some students acquiring detail and speed commensurately.

3 CHAINING

A progressive method is to perform only the beginning stage of a measurement and practice this component before moving on to the next one. This is referred to as '*chaining*', and represents how a new piece of music may be learned in practice, with one section being mastered before the next is attempted. In terms of a skinfold the sequence may be initiated by approaching the skin surface with finger and thumb at 90° to it (Figure 1.1A). Next the finger and thumb are placed in the correct orientation on the landmark (Figure 1.1B). These steps are mastered completely before proceeding further. Once accomplished, the student proceeds to palpate tissue, raising a fold with parallel sides (Figure 1.1B) and then applies the calipers one centimetre away from the inferior edge of the index finger of the left hand and at mid-fingernail depth (Figure 1.1C). The sequence proceeds adding each stage until complete. Many measurements have as many as 10 or so discrete stages, if approaching and orienting the subject are considered. Once all the stages are in place, the timing can be developed, to become a fluid movement. This approach may be useful for subjects who have little or no experience of making anthropometric measurements.

4 SHAPING

This is an alternative approach, where the skill is acquired in a gross form, and modified in a series of stages. Considerations such as subject position, measurer position, orientation and precise timing can be included, as the measurement is perfected. Unlike the previous two methods, this approach does not substantially alter the total time requirement for one measurement. Rather, the consistency within this 'real-time' approach is enhanced as attention to detail increases. Use of digital movie files which can be replayed in a loop, paused or slowed down can enable learning to progress in a more self-directed way, where the individual relates sequences, positions and timing to other key teaching points. This 'mapping' of learning outcomes can even take the form of participants making annotations over a movie clip, effectively writing their own storyboard, which can serve as an *aide-mémoire* to future practice. As students become more fluent, they can compare one another's measurements with the movie clip of the recommended technique. Students with access to a video camera can practise by making movie clips of measuring which can be later reviewed to underscore the learning process. Shaping is an appropriate technique for subjects with some previous experience of making anthropometric measurements.

5 FEEDBACK AND FAULT CORRECTION

With intrinsic feedback, the perception of 'how the movement should feel' can only arise as a consequence of making large numbers of measurements on as many different individuals as possible. In some individuals, landmarks will be difficult to locate, and in others, adipose tissue will adhere to the underlying muscle fascia, making a skinfold very difficult to raise. It is important that students get some exposure to more problematic subjects for measuring if they are to become effective practitioners.

In terms of extrinsic feedback, careful consideration is necessary to inform the process of fault correction. Inappropriate teaching here fosters a dependency on the feedback from the instructor. Common, yet inappropriate is an over-reliance on agreement in scores measured by the instructor and learner. This is particularly destructive in terms of learning with rank novices, who may mistakenly assume close agreement in scores means their technique must have been appropriate. One positive suggestion for teaching skinfolds is to obscure the caliper dial with by a paper cover, until the technique is more consistent. Feedback needs to be delivered in an appropriate way, so maximise the understanding of the learner. In this respect, the '*you-I-you*' approach can be effective, where the perspective of the learner comes first and last, with the perspective of the instructor in between. For example: '*You were standing a long way away from the subject; I'm not sure I could get my head perpendicular to the dial from that position; You could ask yourself next time, how does my body need to be oriented to make sure my line of sight ends up being at 90° to the dial?*'

Focusing feedback on one aspect at a time may avoid confusing the learner who is slow to embrace the complexity of measuring, and encourages further practice where alternative criteria can be considered next. Consideration needs to be given to how long to practise one measurement, before discomfort of the subject or boredom of the learner can undermine the learning process. One solution is to require subjects to perform a series of three or four measurements, thereby acquiring the skills of moving between the individual measurements easily and quickly. (This can enable measurements to be made with greater confidence and releases more time for landmarking or other measurements.) Thus, practice can focus on different aspects in turn, such as body position, hand position, caliper location, timing etc. Above all, it is important for the instructor to remain *positive*, to be *specific*, and to keep comments *concise* and *relevant*.

6 MAXIMUM GROUP ACTIVITY

Attention needs to be paid to the relative independence given to learners, so they can develop skills without becoming reliant on an instructor's intervention. Once the basic skill has been demonstrated, maximum practice is necessary to provide practical experience which complements the theoretical underpinning. This is easiest in anthropometry in a group of three individuals, each taking turns to be measurer, recorder and subject, completing a series of measurements. Groups can then pursue different learning pathways as outlined above. In addition, self-rating or peer-rating can occur with respect to location, orientation, timing, body position etc. for each measurement. As with most skills, the role of the instructor alters as novices gain experience. Responsibility for progression through the learning process can be steadily transferred from the instructor to the student. This will occur at different rates in different individuals, and again, a flexible approach is important. For instance self-directed learning assigned to one group, enables a more labour-intensive fault correction by the instructor in another.

7 SUGGESTIONS FOR STUDENTS OF DIFFERENT LEARNING STYLES

While some students will be self-motivated and confident, others are likely to be shy and reticent. In order to deliver effective teaching to a group containing both, a flexible approach is essential. While independent learners can proceed with limited instructions, expectant or dependant learners wait for opportunities to be developed for them. In any class, both categories are likely to be present, and vigilance early in the instructional session with regard to who requires more didactic teaching is prudent. Failure to do this may mean that the more reticent students accomplish very little during a lesson. However one pitfall of this approach is that the independent learners may be proceeding rapidly, but reinforcing errors which have gone undetected. Solutions to these types of teaching challenge can involve 'opt-in' teaching clinics for methods using only part of the class. Self-paced learning may make use of prepared video clips of teaching material which can be made available in a separate area. Additionally, live video filming of measurements can be performed, and are used widely in sports coaching with very positive results. Students can elect to film or be filmed, and review the video feedback, perhaps comparing against one another, or some pre-prepared material.

In larger classes, teaching ratios may present an additional challenge to learning. However, one valuable contribution to knowledge about effective teaching was recently trialled involving peer-mediated instruction which reinforces attention to detail, which can specifically be problematic in large-group teaching (Dollman, 2005). This involves the instructor dividing the larger class into groups of three or four students and teaching the first of these groups one measurement. After a fixed time interval (for instance, ten minutes), this group passes this teaching onto the next group, and the technique is passed round all the groups, the last of which demonstrate it back to the instructor. (Other tasks such as landmarking or calibrating equipment can be performed by groups awaiting instruction.) Departures from correct technique can be noted and commented on and included in a fuller class discussion after all techniques have been taught. This has been shown to be effective in terms of teaching time, and shows signs of matching or exceeding the accuracy of instructor-centred learning for some measurements.

These principles apply equally to all measurements in anthropometry. For instance, when making girth measurements, separate checks of tape position on the limb segment, orientation (perpendicular to long axis of body segment) and tension (not producing indentation, and spanning concavities), are much more important than the actual results, and represent key teaching points. For segment lengths, placing the segmometer or wide-spreading caliper on the landmarks before measurement can involve a sequence such as '*place – place – check – check – measure*'. For skeletal breadths, general principles may apply such as '*locate generally and then specifically*' and so on. While such principles may not change, the method of teaching delivery can be flexible according to the group size, learning stage and the preferred style of the instructor.

8 CONCLUSION

There are several potential successful routes to learning a practical skill such as making anthropometric measurements. Instructors are of most use to their students if they can deliver a variety of approaches according to different needs, experience levels and learning styles. While instruction in anthropometry has spanned several centuries, approaches to maximising the effectiveness of the learner can benefit from methods used in other practical disciplines, and appropriate use of newer technologies.

REFERENCES

British Association of Ski Instructors, 1997, *Alpine Manual CD*. London: Bell Media Group Ltd.

Dollman, J., 2005, A new peer instruction method for teaching practical skills in the health sciences: an evaluation of the 'Learning Trail'. *Advances in Health Sciences Education* **10**, 125–132.

Durnin, J., 1997, Skinfold thickness measurement Response to letter by Stewart, A. and Eston, R. *Brit. J. Nutr.* **78**, 1040–1042.

Fitts, P.M. and Posner, M.I., 1967, *Human Performance*. Belmont CA: Brooks/ Cole.

International Society for the Advancement of Kinanthropometry (ISAK), 2001, *International Standards for Anthropometric Assessment*. Underdale, SA, Australia.

Kerr, L., Wilkerson, S., Brandy, W.D. and Ishee, J., 1994, Reliability and validity of skinfold measurements of trained versus untrained testers. *Isokinetics and Exercise Science* **4**, 137–140.

Marfell-Jones, M., 2001, The value of the skinfold – background, assumptions, cautions and recommendations on taking and interpreting skinfold measurements. *Proceedings of the Seoul International Sports Science Congress*, pp. 313–323.

Marfell-Jones, M.J., Olds, T., Stewart, A.D. and Carter, L., 2006, *International Standards for Anthropometric Assessment*. International Society for the Advancement of Kinanthropometry (ISAK), Potchefstroom, South Africa.

Wang, J., Thornton., J.C., Kolesnik, S. and Pierson, R.N. Jr., 2000, Anthropometry in body composition. An overview. *Annals of the New York Academy of Sciences* **904**, 317–326.

Webster, J.M., 2002, Skinfold measurement: assessment of the reliability of health/fitness instructors. Eugene, OR: Microform Publications, University of Oregon. http://kinpubs.uoregon.edu/.

World Health Organization, 1989, Measuring Obesity – Classification and Description of Anthropometric Data. *Report on a WHO Consultation on the Epidemiology of Obesity*. National Food and Nutrition Institute, Warsaw.

CHAPTER TWO

Influence of maturation on morphology, food ingestion and motor performance variability of Lisbon children aged between 7 to 8 years

I. Fragoso, F. Vieira, C. Barrigas, F. Baptista, P. Teixeira, H. Santa-Clara, P. Mil-Homens and L. Sardinha

Faculty of Human Kinetics, Technical University of Lisbon, Portugal

1 INTRODUCTION

Although growth is basically driven by the genotype, the growth process employs nutrients as its main source of energy to regulate and restore the body through metabolic and neuroendocrine mechanisms, which can be amplified by physical activity (PA). Daily energy consumption is used to produce voluntary movements (physical activity), in the growth process, in reproduction, in the maintenance of the basal metabolism and in food digestion. Maintenance requirements are dependent on morphology; maturation and activity and growth requirements are dependent on the rate of new cells synthesis. All excess consumed energy is converted into fat and stored in the adipose cells (Cordain *et al.*, 1998).

Nutritional status results from the balance between consumption and expenditure and depends on an individual's physical, social, cultural, and economic environment (Sinclair and Dangerfield, 1998; Vasconcelos, 1993; Krause and Mahan, 1998).

The effect of nutrition and physical activity is such that under-nourished and overweight children, as well as active and inactive children, show differences in both growth and maturation. Although children are the main victims of adverse social conditions, the effects of deficient nutrition in girls seem to be less evident than in boys (Tanner, 1962; Vieira *et al.*, 2002).

The relationship between nutrition and growth was very clear during the First and Second World Wars (1914–1918 and 1939–1945). During those periods the scarcity of food resulted in growth delay among the infantile population due to the suppression of the anabolic effect of growth hormone (GH) (Bogin, 2001). The growth differences between rural and urban populations, after the war, were a good example of GH response to energy consumption. Higher socio-economic level individuals showed greater caloric intakes and were taller, as the result of a catch-up growth process (Sinclair and Dangerfield, 1998).

Deficient or good nutrition is not only a problem of energy balance, but also a problem of diet characteristics. Recently, Neumann *et al.* (2005) showed that children who ate meat had greater increases in slopes for Raven's arithmetic scores, spent more time on high expenditure activities and showed greater frequency of initiative and leadership behaviours. On the other hand, children who predominately had plant-based diets (traditionally regarded as a key source of micronutrients), often had a small consumption of cellular animal protein foods and when protein ingestion is reduced the micronutrient content and its bioavailability is also reduced and as a result the child's growth process is frequently compromised (Gibson, 2005). So, although vegetable ingestion is very important, particularly during growth, children must eat 12% to 14% of animal protein, not only because it helps micronutrient bioavailability, but also because it promotes physical and mental activity and has a pervasive and important role in growth control and in the maintenance of human lean body mass (Borer, 1995).

During the period 1961–2001, Elmadfa and Welchselbaum (2005) observed a notable increase in the proportion of fat intake, considering the total energy supply, and a decrease in the proportion of carbohydrates. Additionally, they observed that European countries had vegetable and fruit intake below recommendations (FAO/WHO/UNU, 2004) and boys consumed less fruit and vegetables than girls (Fragoso and Vieira, 2004; Elmadfa and Welchselbaum, 2005). One of the main consequences of a high-fat diet is a population's weight increases, especially amongst children.

Worldwide, at least 20 million children under 5 years of age are estimated to be overweight (Nishida, 2005). However, there is no reliable global estimate of overweight and obesity in school-age children and adolescents due to the lack of international reference populations. The main determinants of increased childhood obesity are unhealthy diet patterns (increased consumption of saturated fats and simple sugars) and reduced physical activity, both influenced by rapid societal and environmental changes (Nishida, 2005; Parizkova, 2005).

To maintain appropriate energy balance, it is imperative to do physical activity. Physical activity can be defined as any movement, performed by the skeletal muscles, which produces an increase from basal metabolism. This definition includes all kind of activities related with professional performance, domestic tasks, leisure activities, organized sport and many other tasks that can influence the total daily energy expenditure (Bouchard and Shepard, 1994; Philippaerts *et al.*, 1999). Physical activity can be also defined as any activity that increases heart rate and makes people get out of breath (Alexander *et al.*, 2003). Regular physical activity is associated with both immediate and long term health benefits.

The reasons to engage in physical activity are numerous and include weight control, improved muscularity, lower blood pressure, improving cardiorespiratory function and increasing psychological and mental well-being (sheer enjoyment, increasing the sense of achievement or ability, "feeling better"). Although there has been a number of studies carried out on this subject, none was able to establish either the minimum amount of physical exercise necessary to attain a "healthful and harmonious development" or identify the effect of a more intense activity, like training, on a child's development.

Active young people are more likely to report better general health. However, a large cross-sectional observational study in 9 to 10 year old children from four countries in Europe showed that PA explained less than 1% of the variation observed in body fatness and, in a sample of 17-year-old adolescent boys, part of the Stockholm Weight Development Study (Ekelund *et al.*, 2005), only 4%. In fact, boys were significantly more active than girls and were more likely to do, during a week, 60 minutes of moderate to vigorous activity as presented in the WHO Collaborative Cross-National Study – Health Behaviour in School-Aged Children (HBSC) (Alexander *et al.*, 2003). Such results suggest that factors such as food intake and especially the consumption of energy-dense food and sugar-sweetened drinks may play an important role in the development of overweight and obesity, especially in girls.

Clearly physical activity alone is unlikely to adequately counteract a poor diet. It would take between 1 and 2 h of extremely vigorous activity to counteract a single large-sized ($\geq$785 kcal) children's meal at a fast food restaurant (Styne, 2005). Although the above studies failed to prove the relation between activity and the increase of body fat, other studies (Borer, 1995; Fragoso and Vieira, 2004) showed that physical activity patterns established in childhood were reflected later in life. In short, active young people were more likely to be active during adulthood and in addition, a range of psychological, social and environment factors were significantly associated with diaried physical activity and may positively influence general health behaviour, especially nutritional behaviour – so important as already described.

Some recent studies have shown the relation between certain type of activities and nutritional behaviours. Many times, during sedentary activities, individuals eat snacks and other foods dense in calories, but with very little nutritional value, e.g. candies, fries and soft drinks (Fragoso and Vieira, 2004; Utter *et al.*, 2003). Time spent on school work tasks and on reading, however, was positively associated with more healthful food ingestion, such as fruits and vegetables (Utter *et al.*, 2003). There is a growing concern over the small and irregular amounts of physical activity reported by young people. Wang (2005) has shown that only 26% of boys reported more than 20 minutes of vigorous-moderate exercise in 5 days, across a 7-day period. Ideally, all young people should participate in physical activity, of at least a moderate intensity, for one hour each day for not less than five days a week (Alexander *et al.*, 2003).

The energy recommendations (FAO/WHO/UNU, 2004) for children of all ages are based on measurement and estimates of total daily energy expenditure (dependent on morphology, maturation and activity), to which are added the energy needs for tissue deposition related to growth in case of infants and children (Uauy, 2005). The available data show a weaker, but significant, interrelation between physical activity and maturation during childhood and a strong association during adolescence, especially when speaking about strength. In addition, the amount of fat-free mass measured through mid-thigh girths or through anthropometric equations, seems to be a good measure of a child's maturational status (Capela *et al.*, 2005; Fragoso *et al.*, 2004a,b, 2005). Thus, if energy requirements for maintenance are related to physical activity and morphology and if children with different maturation levels have different morphologic dimensions, characteristic of their growth process, it is expected also to find an interrelation between maturation and energy intake. Therefore, the objective of this study was to evaluate whether maturation variability is an important variable to distinguish groups with different energy requirements.

2 METHODS

The sample was composed of 86 children (45 girls and 41 boys) aged between 7.8 and 8.9 years. The anthropometric measures obtained were: height and weight; triceps, subscapular, biceps, iliac crest, abdominal, front thigh and medial calf skinfolds; biepicondylar humerus and biepicondylar femur breadths; and arm relaxed, arm flexed and tensed, mid-thigh and medial calf girths. All variables were measured according to the ISAK (2001) protocols. In addition, three non-ISAK skinfolds were taken: thoracic (immediately after the chest sulcus in the direction of mamilla); chest (taken at the mid-point between the anterior axillary fold and the mamilla in the boys and one third of the distance between the anterior axillary fold and the mamilla in the girls); and mid-axilla (taken on the ilio-axillary line at the level of the xiphoidale landmark). The sum of skinfolds (TotalSK) was used as an indicator of total fat, and the sums of limb skinfolds (LimbSK) and of trunk skinfolds (TrunkSK) were used as indicators of fat distribution. Body composition was estimated through Slaughter *et al.* (1988) and Lohman (1986) equations. Percent body fat was calculated from the mean value of the two equations.

From the literature, boys seemed to be more active than girls when evaluated through time expenditure in organized and leisure activities obtained by self-report methods (Strauss *et al.*, 2001; Kemper *et al.*, 1999; Matos *et al.*, 2003; Mota and Silva, 1999; Center for Disease Control and Prevention, 2003; Vilhjalmsson and Kristjansdottir, 2003; Fragoso and Vieira, 2004). However, self-reported activity might contain errors due to the misperceptions of the respondents (Styne, 2005; Ekelund *et al.*, 2004). To minimize this problem, the total amount of time (minutes) spent on physical activity was objectively recorded each day with a uni-axial accelerometer (Manufacturing Technology Incorporated, Fl, USA, model WAM 6471, formerly known as the CSA activity monitor). The accelerometer was secured on the right hip, using an elastic belt. The subjects were asked to wear the accelerometer during the daytime for five week days and two weekend days, except during water activities. Activity data were analyzed and processed using a special written macro based on Microsoft Excel. The output from the macro included cumulated time spent at PA of moderate and high intensity levels, as defined by Trost *et al.* (2002). Further, the total amount of PA time (minutes) registered each day was recorded. The total amount of PA was expressed as total counts divided by registered time, i.e. counts per minutes. All activity data were averaged over the 7-day period.

The primary physical activity outcome variable was total counts divided by the registered time. Secondary outcome variables were time spent at sedentary activity, time spent at light intensity of physical activity, time spent at moderate intensity of physical activity and time spent at vigorous intensity of physical activity. Age-specific cut-off values for moderate and vigorous intensity were adopted from Trost *et al.* (2002). An arbitrary cut-off value, previously used by Ekelund *et al.* (2004) in adolescents, of ≤100 counts.min^{-1} was used as an indicator of sedentary behaviour. Light intensity of physical activity was defined as the time accumulated >100 counts.min^{-1} and less than the specific cut-off point corresponding to moderate intensity of physical activity for each age.

We used a semi-quantitative questionnaire to assess the frequency of food ingestion of a wide set of typical Portuguese foods. The Food Frequency Questionnaire used in this study (Moreira *et al.*, 2003), included 89 habitual Portuguese foods. This questionnaire was compared with a 3-Day Dietary Food Record, showing an adequate validity (Moreira *et al.*, 2003). The response to this questionnaire was achieved via a child's parent or guardian interview. For each listed food product, the child's parent or guardian chose the amount (a clearly described pre-established portion) regularly ingested by their offspring. The quantitative data was then introduced into a computerized database for determination of the relative consumption of the different nutrients.

The aerobic capacity was assessed by the PACER test (Progressive aerobic cardiovascular endurance run) as described in the test administration manual of Fitnessgram 6.0 program (Welk and Meredith, 1999). In this test, a child must move 20 meters, from one line to another, then reverse direction, in accordance with the cadence defined by a music-CD. The CD contains 21 effort levels (one per minute) of increasing velocity, thus in the beginning the child walks then gradually starts to run. All the children were assessed in the same conditions, but not necessarily on the same occasion. All the tests were performed in the same Gym, on the same day, between 11:00 and 13:00 hours. Each child had a trained adult to record the lap number and to control the test performance. All the children had a warm-up before the test and they were trained for the 20-meter shuttle run with the CD beep music version. The VO$_2$ max was calculated using the Oliveira (1996) equation: VO$_2$ max = 28.1 + 0.35 laps n°.

We also measured muscle capacity using two strength motor tasks – the long stretch-shortening cycle evaluated with a counter-movement jump and a handgrip strength test (Bosco and Komi, 1979; Schmidtbleicher, 1985).

Maturational measures consisted of a skeletal age evaluation obtained through radiographs of left hand and wrist. The maturity ratings were done by two examiners trained at the Faculty of Human Kinetics of Lisbon who were blinded to the chronological age of the subjects. Thirteen bones were rated by comparing the ossification stage of each bone according to the Tanner-Whitehouse III (TW3) method (Tanner *et al.*, 2001).

The sample was divided into three different groups based on identified maturation level. The low-maturation group children showed differences between bone age and chronological age inferior or equal to –0.59 years; the medium-maturation group children showed differences between bone age and chronological age greater than –0.59 years, but less than 1.1 years; and the high-maturation group children showed differences between bone age and chronological age equal or superior to 1.1 years.

The descriptive statistics, the one sample T-test and the multivariate anova were generated by SPSS version 12 for Windows. Gender was the covariate utilized. In all statistical tests, the level of significance was set at p ≤ 0.05.

3 RESULTS AND DISCUSSION

To see if the used sample was representative of the general Lisbon population, we compared the morphological values of boys and girls, with a Lisbon reference sample of 366 boys and 353 girls of 8 years old (Fragoso and Vieira, in press 2007).

Table 2.1 Means and Standard Deviations of Anthropometric and Body Composition measures of boys and girls.

	Girls (N = 45)		Boys (N = 41)	
	Mean	SD	Mean	SD
Chronological Age	8.41	0.29	8.40	0.28
Bone Age	8.58	1.34	8.71	1.09
Weight (kg)	29.88	5.49	31.61	7.66
Stature (cm)	131.32	5.20	131.88	7.08
Biceps sf (mm)	7.81	3.68	7.22	3.62
Triceps sf (mm)	13.02	4.16	13.24*	5.33
Front thigh sf (mm)	19.28	5.37	18.11	6.66
Medial calf sf (mm)	12.62	4.31	12.63	5.03
Subscapular sf (mm)	8.00*	4.34	7.58	3.34
Thoracic sf (mm)	7.78	4.02	8.76	4.32
Chest sf (mm)	8.31	4.16	8.34	4.65
Mid-axilla sf (mm)	7.42	4.34	7.49	3.96
Abdominal sf (mm)	11.36	5.61	11.90	7.50
Iliac crest sf (mm)	11.39	5.96	10.90*	5.97
Total SK	106.99	41.14	106.18	46.49
Trunk SK	54.26	26.48	54.97	28.17
Limb SK	52.73	16.09	51.21	19.88
Arm Relax gth (cm)	20.06*	2.62	20.65	3.34
Arm Tensed gth (cm)	20.79	2.43	21.50	3.01
Mid Thigh gth (cm)	37.52	4.04	38.33	5.53
Medial Calf gth (cm)	27.20	2.38	27.87	3.28
Bi Epi. Humerus (cm)**	5.16*	0.41	5.35	0.43
Bi Epi. Femur (cm)**	7.65*	0.48	8.06	0.64
Fat %	19.80	5.64	18.84	6.41

* Significant differences between our sample and the reference sample
** Significant differences between boys and girls

The morphological characteristics of this sample were similar to those of the Lisbon children population with only some exceptions: the triceps skinfold (13.2 vs 11.1 mm) and iliac crest skinfold (10.9 vs 8.3 mm) for boys, and subscapular skinfold (8.0 vs 10 mm), relaxed arm girth (20.1 vs 21.0 cm), bi-epicondylar humerus breadth (5.2 vs 5.0 cm) and bi-epicondylar femur breadths (7.7 versus 7.5 cm) for girls (Table 2.1).

There was a great similarity between the boys' and girls' chronological ages in both samples. However, the boys in the test sample had a slightly superior bone age, had significantly more triceps and iliac crest adiposity and more relaxed arm girth. The only significant differences between both genders were for bi-epicondylar humerus (p = 0.032) and femur (p = 0.001) breadths (Table 2.1).

As expected from Ekelund *et al.* (2004), the fraction of time spent at vigorous PA was significantly higher in boys (10.8 vs 5.6 min.d^{-1}). (See Table 2.2.) However, type, intensity and frequency of activities were not identical in boys and girls. Girls did much more domestic activities and moderate activities than boys and boys more vigorous activities as has been identified elsewhere (Chadwick and Fragoso, in Press 2007). So, gender differences were not a concern during this stage, but the small amount of vigorous activity shown by both groups was.

Table 2.2 Means and Standard Deviations of Physical Activity and Food Ingestion measures of boys and girls.

	Girls (N = 44)		Boys (N = 40)	
	Mean	SD	Mean	SD
Physical Activity (PA)				
Vigorous PA (min.d^{-1})**	5.61	6.00	10.75	6.92
Right HG (kg)	13.55	3.21	14.78	2.76
Left HG (kg)	12.48	3.45	13.40	2.81
CM jump (cm)	23.07	3.80	21.51	3.56
Shuttle-Run (laps)	23.12	9.31	27.85	12.45
VO$_2$max (ml/kg/min)	36.19	3.26	37.60	4.57
Food Ingestion				
Calories (kcal)**	2072.98	739.58	2676.33	1451.43
Protein (g)	91.06	36.31	109.09	53.92
Carbohydrates (g)**	272.76	107.04	359.12	199.45
Total Fat (g)**	73.17	26.76	95.48	55.95

** Significant differences between boys and girls (p = 0.02 for Calories, p = 0.01 for carbohydrates and p = 0.02 for total fat)

Other than vigorous PA, there were no differences between physical activity parameters between the genders (Table 2.2). The total caloric ingestion as well as daily protein, carbohydrates and fat consumptions were higher in boys, although the only non-significant gender difference was related to daily protein intake (p = 0.07).

The results show that, in general, there are significant differences between the median and the high maturation groups for all anthropometric (Table 2.3) and body composition variables (Table 2.4).

Table 2.3 Comparison of Anthropometric measures between maturational groups without sex effects

	Mat. Groups	Desc. Statistics Mean	SD	Mean Est.	Pairwise Comp. Groups	Sig.
Stature	1 (Low)	128.06	4.68	127.94	1 vs 2	0.14
	2 (Median)	130.62	5.73*	130.66	1 vs 3	0.00
	3 (High)	135.71	4.84*	135.97	2 vs 3	0.00
Relaxed Arm Girth	1 (Low)	18.56	2.25	18.26	1 vs 2	0.07
	2 (Median)	19.97	2.60*	19.86	1 vs 3	0.00
	3 (High)	22.32	2.87*	22.49	2 vs 3	0.00
Arm Girth (tensed and flexed)	1 (Low)	19.69	2.34	19.32	1 vs 2	0.08
	2 (Median)	20.88	2.44*	20.75	1 vs 3	0.00
	3 (High)	22.72	2.57*	22.88	2 vs 3	0.00
Mid-thigh Girth	1 (Low)	34.58	3.44*	34.22	1 vs 2	0.03
	2 (Median)	37.56	4.30*	37.49	1 vs 3	0.00
	3 (High)	41.18	4.86*	41.42	2 vs 3	0.00
Med Calf Girth	1 (Low)	25.58	1.88*	25.33	1 vs 2	0.03
	2 (Median)	27.29	2.57*	27.20	1 vs 3	0.00
	3 (High)	29.49	2.89*	29.64	2 vs 3	0.00
Bi Epi. Humerus**	1 (Low)	4.96	0.24*	4.96	1 vs 2	0.02
	2 (Median)	5.28	0.44*	5.27	1 vs 3	0.00
	3 (High)	5.45	0.35*	5.50	2 vs 3	0.05
Bi Epi. Femur**	1 (Low)	7.41	0.38*	7.41	1 vs 2	0.02
	2 (Median)	7.83	0.50*	7.79	1 vs 3	0.00
	3 (High)	8.31	0.61*	8.37	2 vs 3	0.00

** Significant gender differences
*Significant differences between maturational groups

All measures increased along the three maturation levels, but only the weight (25.89 kg, 29.81 kg, 36.09 kg), midthigh girth (34.58 cm, 37.56 cm, 41.18 cm), medial calf girth (25.59 cm, 27.29 cm, 29.49 cm), biepicondylar humerus (4.96 cm, 5.28 cm, 5.45 cm) and biepicondylar femur (7.41 cm, 7.83 cm, 8.31 cm) values were significantly different across all maturation groups. It seems that maturation level is particulary important as far as robustness development is concerned (Capela *et al.*, 2005; Fragoso *et al.*, 2004a, 2005).

Table 2.4 Comparison of Body Composition measures between maturational groups without sex effects.

	Mat. Groups	Desc. Statistics		Mean Est.	Pairwise Comp.	
		Mean	SD		Groups	Sig.
Weight	1 (Low)	25.89*	3.65	25.50	1 vs 2	0.03
	2 (Median)	29.81*	5.68	29.60	1 vs 3	0.00
	3 (High)	36.09*	6.32	36.51	2 vs 3	0.00
Total SK	1 (Low)	89.57	37.97	82.97	1 vs 2	0.30
	2 (Median)	97.54*	37.80	96.77	1 vs 3	0.00
	3 (High)	135.26*	43.50	137.38	2 vs 3	0.00
Trunk SK	1 (Low)	43.57	21.81	40.57	1 vs 2	0.32
	2 (Median)	49.81*	24.15	49.01	1 vs 3	0.00
	3 (High)	71.76*	28.92	72.83	2 vs 3	0.00
Limb SK	1 (Low)	46.00	17.87	42.40	1 vs 2	0.32
	2 (Median)	47.73*	14.74	47.76	1 vs 3	0.00
	3 (High)	63.50*	17.16	64.55	2 vs 3	0.00
Fat %	1 (Low)	16.66	5.78	15.58	1 vs 2	0.15
	2 (Median)	18.16*	5.19	18.17	1 vs 3	0.00
	3 (High)	23.15*	5.27	23.31	2 vs 3	0.00

*Significant differences between maturational groups

Body fat measures and the sum of skinfolds were significantly different between the median and high groups of maturation. These results were in accordance with what Bogin (2001), Sinclair and Dangerfield (1998) and Tanner (1962) have called secular trend effects. Part of secular height differences, during childhood and adolescence, are due to children's early maturation and related to the increase of weight and fat mass during childhood. In our view, we are dealing with a problem of positive energy balance and human survival. It is much better and less harmful to our bodies to grow faster than to be obese.

Although maturation levels explained the variation of all morphologic variables, it was not related to caloric ingestion, since the low maturation group had higher total calories consumption, proteins, carbohydrates and total fat consumption. These results were not significantly different for all maturation groups (Table 2.5). This means that our sample had independent energy consumption and maturation levels. When nutrient energy is limited due to compulsory forms of exercise, to lack of food ingestion or to a deficit of certain macronutrients, growth and maturation are affected and most of the time delayed. Inversely, when nutrient energy is not limited, growth and maturation are highly independent processes (Malina, 1989). Still, when positive energy balance is considered, the body uses its extra energy to grow faster and reduce the probability of becoming fat.

Table 2.5 Comparison of food ingestion measures between maturational groups without sex effects.

	Mat. Groups	Desc. Statistics		Mean Est.	Pairwise Comp.	
		Mean	SD		Groups	Sig.
Calories	1 (Low)	2484.18	1616.47	2519.32	1 vs 2	0.42
	2 (Median)	2311.72	896.28	2287.37	1 vs 3	0.50
	3 (High)	2322.70	1184.98	2308.28	2 vs 3	0.94
Protein	1 (Low)	110.10	60.53	110.39	1 vs 2	0.16
	2 (Median)	95.26	38.22	94.19	1 vs 3	0.31
	3 (High)	97.90	45.29	67.62	2 vs 3	0.77
Carbohyd.	1 (Low)	315.69	211.56	323.38	1 vs 2	0.73
	2 (Median)	311.93	127.60	309.51	1 vs 3	0.75
	3 (High)	314.22	180.36	309.71	2 vs 3	0.99
Total Fat	1 (Low)	92.06	66.20	92.47	1 vs 2	0.27
	2 (Median)	81.51	33.88	80.40	1 vs 3	0.30
	3 (High)	79.61	36.32	80.13	2 vs 3	0.98

Table 2.6 Comparison of Physical Activity measures between maturational groups without sex effects.

	Mat. Groups	Desc. Statistics		Mean. Est.	Pairwise Comp.	
		Mean	SD		Group	Sig.
Vig. Act*	1 (Low)	9.28	8.17	8.92	1 vs 2	0.50
	2 (Median)	8.33	6.68	7.45	1 vs 3	0.85
	3 (High)	7.86	7.05	8.44	2 vs 3	0.60
Right HG	1 (Low)	12.79	2.75	12.93	1 vs 2	0.23
	2 (Median)	14.14*	2.75	14.04	1 vs 3	0.01
	3 (High)	15.41	2.43	15.55	2 vs 3	0.06
Left HG	1 (Low)	11.79	3.70	11.54	1 vs 2	0.15
	2 (Median)	13.12	2.72	12.97	1 vs 3	0.07
	3 (High)	13.59	2.76	13.63	2 vs 3	0.45
CM jump	1 (Low)	23.11	3.41	22.75	1 vs 2	0.48
	2 (Median)	21.81	3.62	21.87	1 vs 3	0.42
	3 (High)	21.90	4.25	21.59	2 vs 3	0.80
$VO_{2\ max}$	1 (Low)	37.63	4.08	38.52	1 vs 2	0.22
	2 (Median)	37.04	4.14	36.85	1 vs 3	0.17
	3 (High)	36.38	3.96	36.34	2 vs 3	0.67

*Significant differences between maturational groups

Ekelund *et al.* (2004) reported that sexual maturity had no significant effect on PA components of children aged 9 to 10 years old. Our data confirmed this result since, in our study, maturation level, defined by the difference between bone age and chronological age, had no influence on motor performance (Table 2.6). However some consideration can be given: (1) the less-mature children were not only the more active ones, but also the ones that had higher values for counter-movement jump and VO2 max. This fact can be related to their morphologic characteristics (less mature children were lighter and smaller than the other children); (2) the more mature children had more right and left handgrip strength which confirmed the relation between bone age and handgrip strength as described by Fragoso *et al.* (2004). We can speculate that caloric consumption was not only related to vigorous physical activity, but also to spontaneous physical activity. However, as the latter was not discriminated in this study, this remains conjecture.

The fact that low maturation levels were more active is not potentially of concern because naturally, mobility and spontaneous activity decrease during childhood. What is of real concern, however, is the tiny amount and irregularity of physical activity across all maturation levels in the sample studied, who presented only 7.86–9.28 min of vigorous activity per day. This is a very modest value considering that Portuguese children watch aproximately 16 hours a week of TV (Fragoso and Vieira, 2004b). Wang (2005) had similar findings, showing that only 26% of boys reported more than 20 minutes of vigorous-moderate exercise in 5 days.

It seems that, in general, the maturation level explains the variation of all morphologic variables. However, only the weight and limb measures presented significant differences between all maturation groups. More mature children present greater right handgrip strength (12.92 kg, 14.03 kg, 15.55 kg). The results showed that maturation was not related to aerobic and counter movement jump performance. The midthigh girth, medial calf girth and bi-epicondylar breadths were good robustness indicators and seemed to reflect the child's maturational status. Total caloric energy intake and time spent in vigorous physical activity were independent of maturation variability. Still, the group with the lower level of maturation ate more calories and spent more time on vigorous activity. Finally, although not significantly, food ingestion (2484.18 kcal, 2311.72 kcal, 2322.70 kcal) seemed to be associated with the time spent in vigorous daily activity (8.92 min.d^{-1}, 7.45 min, 8.44 min).

4 CONCLUSION

Ideally, and in accordance to Alexander *et al.* (2003), all young people should participate in physical activity of at least a moderate intensity for one hour each day, for not less than five days a week. However, regular exercise, in agreement with Ekelund *et al.* (2005), only explained less than 1% of body fatness of a sample of 9 to 10 year old children from four countries in Europe and 4% in a 17-years-old adolescent boy's sample, part of the Stockholm Weight Development Study. So, according to the scientific community, a certain frequency and intensity of exercise can play a pervasive and important role in growth control and in the maintenance of human lean body mass (Borer, 1995) which may play an important role in weight control. Obesity is the result of a rapid societal and environmental changes associated with early maturation, an enormous decrease in non-regular physical activity (children don't help their parents any more in agriculture labor, or in the home), the decrease of muscle-skeletal development and poor eating habits.

Due to the significant amounts of spontaneous activity during the second part of childhood, a decrease in the amount of body fat would be expected. However, over the past few decades, adults have compressed child activity routines, changed children's free time and children have became small adults. They wake up at the same time as the adults do, they stay away from their family all day and when they return home they have to do their homework and play what is possible in modern cities (gameboy, playstation or watch television). In other words, their free time has been reduced and the number of hours spent in sedentary activities increased drastically and radically. As a result, although children demonstrate the capacity for extraordinary mobility during childhood, they have no time to express it (except those who have the opportunity, during a short period of time, to participate in any organized activity). This fact has irreparable consequences because it is contributing to the decrease of muscle growth and to the increase of fatty cell dimension as well as to its hyperplasia.

We need to promote more regular activity as children mature, but above all we have to combine efforts in the sense of promoting physical activity for all children (American Academy of Pediatrics, 2000; WHO, 2003). The difficulty when dealing with obesity is neither the high calorie ingestion and its control nor the lack of physical activity and its promotion. What is really difficult is to control energy balance and its different components. As demonstrated in this study, when nutrition is not limited, maturation is correlated with fat percent, but not with calorie ingestion which makes energy balance during growth a complex issue. Apparently, above a certain amount of fat mass, caloric consumption is not necessary related with maturation (biologic age). Therefore, we presume that there are groups of children with specific energy balance characteristics and needs. In conclusion, to obviate future generations being heavier and less active it is important to: (1) improve all children's lifestyle, teaching them to eat better and organizing their time in a way to offer them free time and appropriate spaces for different and spontaneous (natural) activities; (2) plan special activities for groups of children with particular energy balance characteristics.

5 REFERENCES

Alexander, L., Currie, C. and Todd, J., 2003, Gender matters: Physical activity patterns of schoolchildren in Scotland. In *HBSC – Health Behaviour in School-Aged Children: WHO Collaborative Cross-National Study*. Briefing Paper 3 (October 2003).

American Academy of Pediatrics, 2000, Physical fitness and activity in schools. *Pediatrics*, **105**, 1156–1157.

Bogin, B., 2001, *The Growth of Humanity*. New York: Wiley-Liss.

Borer, K.T., 1995, The effects of exercise on growth. *Sports Medicine*, **20**, 375–397.

Bosco, C. and Komi, P.V., 1979, Potentiation of the mechanical behavior of the human skeletal muscle through prestretching. *Acta Physiologica Scandinavica*, **106**, 467–472.

Bouchard, C. and Shephard, R.J., 1994, Physical activity, fitness and health: The model and key concepts. In *Physical Activity, Fitness, and Health: International Proceedings and Consensus Statement*, edited by Bouchard, C., Shephard, R.J. and Stephans, T. (Champaign, IL: Human Kinetics), pp. 77–88.

Capela, C., Fragoso, I., Vieira, F., Mil-Homens, P., Gomes Pereira, J., Charrua, C., Lourenço, N. and Gonçalves, Z., 2005, Physical performance tests in young soccer players with reference to maturation. In *Sciences and Football V – The Proceedings of the Fifth World Congress on Sciences and Football*, edited by Reilly, T., Cabri, C. and Araújo, D. (London: Routledge. Taylor & Francis Group), pp. 429–433.

Center for Disease Control and Prevention, 2003, Physical activity levels among children aged 9–13 years – United States, 2002. *Morbidity and Mortality Weekly Report*, **52**, 785–788.

Chadwick, D. and Fragoso, I., 2007 in press, Serão as raparigas inactivas? *Proceedings of IPA's XVI World Congress, Lisbon 21st to 25th June 1999.* Lisboa: FMH Edições.

Cordain, L., Gotshall, R.W., Boyd Eaton, S. and Boyd Eaton III, S., 1998, Physical activity, energy expenditure and fitness: An evolutionary perspective. *International Journal of Sports Medicine*, **19**, 328–335.

Ekelund, U., Neovius, M., Linné, Y., Brage, S., Wareham, N.J. and Rossner, S., 2005, Associations between physical activity and fat mass in adolescents: The Stockholm weight development study. *American Journal of Clinical Nutrition*, **81**, 355–360.

Ekelund, U., Sardinha, L.B., Anderssen, S.A., Harro, M., Franks, P.W., Brage, S., Cooper, A.R., Andersen, L.B., Riddoch, C. and Froberg, K., 2004, Associations between objectively assessed physical activity and indicators of body fatness in 9- to 10-y-old European children: A population-based study from 4 distinct regions in Europe (the European Youth Heart Study). *American Journal of Clinical Nutrition*, **80**, 584–590.

Elmadfa, L. and Welchselbaum, E., 2005, The European nutrition and health report 2004. In *Proceedings of the Pre-Congress Safaris: Indigenous People's Food Systems and Nutrition*, p. 19.

FAO/WHO/UNU, 2004, *Vitamin and Mineral Requirements in Human Nutrition.* [on line], available at whqlibdoc.who.int/publications/2004/9241546123.pdf.

Fragoso, I. and Vieira, F., 2004, Actividade física e práticas alimentares de crianças de Lisboa (RAPIL). Consequências obesogénicas. *Revista Portuguesa de Ciências do Desporto*, **4**, 65–67.

Fragoso, I. and Vieira, F, 2007 in press, *Reavaliação Antropométrica da População de Lisboa. Tendência Secular.* Lisboa: Imprensa Nacional.

Fragoso, I., Fortes, M., Vieira, F. and Canto e Castro, L., 2004a, Different maturational levels during adolescence: A methodological problem. In *Kinanthropometry VIII*, edited by Reilly, T. and Marfell-Jones, M. (London: Taylor and Francis), pp. 68–81.

Fragoso, I., Vieira, F., Canto e Castro, L., Júnior, A.O., Capela, C., Oliveira, N., and Barroso, A., 2004b, Maturation and strength of adolescent soccer players. In *Biossocial Approach to Youth Sports*, edited by Coelho e Silva, M. and Malina, R. (Coimbra: Coimbra University Press), pp. 199–208.

Fragoso, I., Vieira, F., Canto e Castro, L., Mil-Homens, P., Capela, C., Oliveira, N., Barroso, A., Veloso, R. and Júnior, A.O., 2005, The importance of chronological and maturation age on strength, resistance and speed performance of soccer players during adolescence. In *Sciences and Football V – The Proceedings of the Fifth World Congress on Sciences and Football*, edited by

Reilly, T., Cabri, C. and Araújo, D. (London: Routledge, Taylor & Francis Group), pp. 465–470.

Gibson, R.S., 2005, Traditional methods for food processing, diet modification, and diversity to increase micronutrient content and bioavailability. In *Proceedings of the Pre-Congress Safaris: Indigenous Peoples's Food Systems and Nutrition*, p. 10.

International Society for the Advancement of Kinanthropometry (ISAK), 2001. *International Standards for Anthropometric Assessment*. (Underdale, SA, Australia.)

Kemper, H.C.G., Post, G.B., Twisk, J.W.R. and van Mechelen, W., 1999, Lifestyle and obesity in adolescence and young adulthood: Results from the Amsterdam Growth and Health Longitudinal Study (AGAHLS). *International Journal of Obesity*, **23** (Suppl. 3), S34–S40.

Krause, M.V. and Mahan, L.K., 1998, *Alimentos, Nutrição e Dietoterapia*. São Paulo: Editora Roca.

Lohman, T.G., 1986, Applicability of body composition techniques and constants for children and youth. *Exercise and Sports Science Review*, **14**, 325–357.

Malina, R.M., 1989, Growth and Maturation: Normal Variation and Effect Training. In *Exercise Science and Sports Medicine*, edited by Gisolfi, C.V. and Lamb, D.R., (Indianapolis, Indiana: Benchmark Press), pp. 223–265.

Matos, M.G. and Equipa do Projecto Aventura Social e Saúde, 2003, *A Saúde dos Adolescentes Portugueses (Quatro anos depois)*. Cruz Quebrada: FMH/Programa de Educação para Todos-Saúde.

Moreira, P., Sampaio, D. and Almeida, M.D.V., 2003, Validade relativa de um questionário de frequência de consumo alimentar através da comparação com um registo alimentar de quatro dias. *Acta Médica Portuguesa*, **16**, 412–420.

Mota, J. and Silva, G., 1999, Adolescent's physical activity: Association with socioeconomic status and parental participation among a Portuguese sample. *Sport, Education and Society*, **4**, 193–199.

Neumann, C.G., Gewa, C.G. and Murphy, S.P., 2005, Meat supplementation improves micronutrients nutrition, growth, and functional outcomes in Kenyan school children. In *Proceedings of the Pre-Congress Safaris: Indigenous Peoples's Food Systems and Nutrition*, p. 88.

Nishida, C., 2005, Childhood obesity – a new challenge for public health nutrition. In *Proceedings of the Pre-Congress Safaris: Indigenous Peoples's Food Systems and Nutrition*, p. 87.

Oliveira, J., 1996, Validação directa do teste de vai-vem 20 m de "Luc-Léger". *Tese de Mestrado*. Lisboa: Faculty of Human Kinetics.

Parizkova, J., 2005, Obesity in women in Eastern Europe. In *Proceedings of the Pre-Congress Safaris: Indigenous Peoples's Food Systems and Nutrition*, p. 87.

Philippaerts, R.M., Lefevre, J., Delvaux, K., Thomis, M., Vanreusel, B., Lysens, R. and Beunen, G., 1999, Associations between daily physical activity and physical fitness in Flemish Males: A cross-sectional analysis. *American Journal of Human Biology*, **11**, 587–597.

Schmidtbleicher, D., 1985, Classification des methods d'entraînement de la Force. *Revue de L' AEFA*, **93**, 29–32.

Sinclair, D. and Dangerfield, P., 1998, *Human Growth After Birth*. New York: Oxford University Press.

Slaughter, M.H., Lohman, T.G., Boileau, R.A., Horswill, C.A., Stillman, R.J. van Loan, M.D. and Bemben, D.A., 1988, Skinfold equation for estimation of body fatness in children and youth. *Human Biology*, **60**, 709–723.

Strauss, R.S., Rodzilsky, D., Burack, G. and Colin, M., 2001, Psychosocial correlates of physical activity in healthy children. *Archives of Pediatrics & Adolescent Medicine*, **155**, 897–902.

Styne, D.M., 2005, Obesity in childhood: What's activity got to do with it? *American Journal of Clinical Nutrition*, **81**, 337–338.

Tanner, J., 1962, *Growth at Adolescence*, 2nd ed. Oxford: Blackwell Scientific Publications.

Tanner, J.M., Healy, M.J.R., Goldstein, H. and Cameron, C., 2001, *Assessment of Skeletal Maturity and Prediction of Adult Height (TW3 Method)*. London: W.B. Saunders.

Trost, S.G., Pate, R.R., Sallis. J.F., Freedson, P.S., Taylor, W.C., Dowda, M. and Sirard, J., 2002, Age and gender differences in objectively measured physical activity in youth. *Medicine & Science in Sports & Exercise*, **34**, 350–355.

Uauy, R., 2005, New norms on energy needs and growth of young children: Implications for the prevention and control of obesity. In *Proceedings of the Pre-Congress Safaris: Indigenous People's Food Systems and Nutrition*, p. 16.

Utter, J., Neumark-Sztainer, D., Jeffery, R. and Story, M., 2003, Couch potatoes or french fries: Are sedentary behaviors associated with body mass index, physical activity, and dietary behaviors among adolescents? *Journal of the American Dietetic Association*, **103**, 1298–1305.

Vasconcelos, F.A.G., 1993, *Avaliação Nutricional de Coletividades*. Santa Catarina: Editora da UFSC.

Vieira, F., Fragoso, I., Ferreira, C., Oliveira, C., Barrigas, C. and Silva, L., 2002, Maturational and nutritional levels of children aged between 6 and 10 years. *Humanbiologia Budapestinensis*, **27**, 143–151.

Vilhjalmsson, R. and Kristjansdottir, G., 2003, Gender differences in physical activity in older children and adolescents: The central role of organized sport. *Social Science and Medicine*, **56**, 363–374.

Wang, Y., 2005, Promise of environmental approaches: A school-based childhood obesity intervention study. In *Proceedings of the Pre-Congress Safaris: Indigenous Peoples's Food Systems and Nutrition,* p. 88.

Welk, G. and Meredith, M., 1999, *Fitness Program of the Cooper Institute for Aerobics Research*. Human Kinetics, Champaign, IL.

WHO, 2003, *WHO Global Strategy on Diet, Physical Activity and Health*. [Online], available at http://www.who.int/hpr/global.strategy.shtml.

Acknowledgements

We are very grateful to the participants and their families who gave their time to this study and to the members of Faculty of Human Kinetics who have participated as anthropometrists.

Acknowledgement is also made to Portugal Institute of Sport (grant 236/2004) and Lilly Portugal Pharmaceutical who supported this study.

CHAPTER THREE

Virtual anthropometry

T. Olds[1], J. Ross[2], P. Blanchonette[3] and D. Stratton[4]

[1] University of South Australia
[2] Defence Health Services, Royal Australian Air Force
[3] Defence Science Technology Organisation, Australia
[4] University of Ballarat, Australia

1 INTRODUCTION

1.1 Secular changes in body size and shape

In most countries, there have been dramatic changes in body size and shape over the last 50–100 years (Meredith, 1976). Figure 3.1 summarizes 30 studies which have tracked the evolution of height in various populations. Within each study, height at various time points was expressed as a percentage of the measured or estimated height in 1990. It appears that the rate of increase in height is accelerating. Most of the increase in height has occurred in the femur. Consequently, sitting height to stature ratios have fallen. These proportionality changes have implications for comfortable seating in aeroplanes, trains, trams and automobiles, and for manoeuvres such as ejection from aircraft.

Shape has also been changing. Waist girth has been increasing at a far greater rate than hip or chest girth. People are also becoming fatter, marked by increasing subcutaneous skinfold thicknesses, and increasing weight and body mass index (BMI). While there are few data on adults, triceps skinfold thickness and estimated percentage body fat in children have been increasing at about 8 per cent per decade since the 1950s, based on collated data from over 500,000 children. The changing ethnic mix of most populations, and the integration of women into and the exclusion of children from wider workplace and operational domains have also changed the effective body size and shape mix. Table 3.1 compares typical dimensions for a 20–50 year old Western male in 1955 and 2005.

height (% of 1990 values)

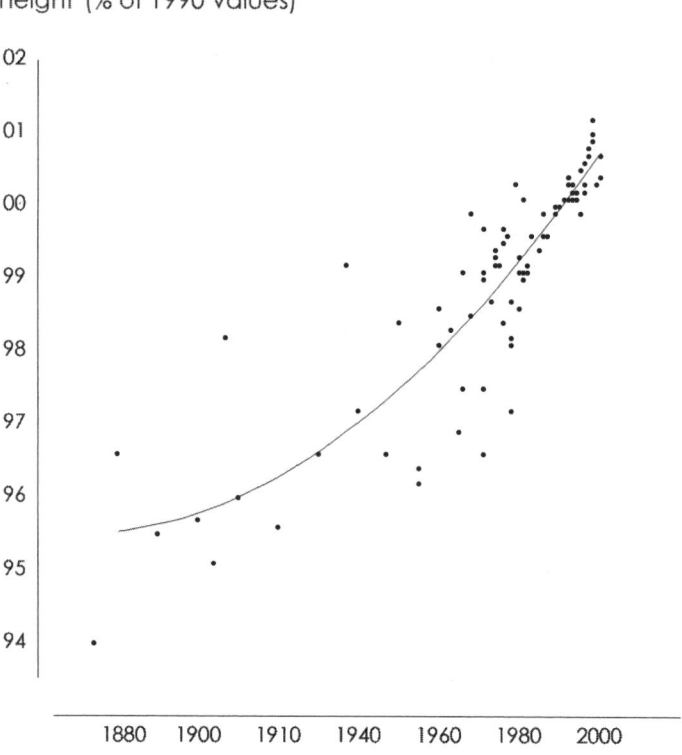

year of measurement

Figure 3.1 Cumulated data from 30 studies showing the evolution of adult height. Heights are expressed as a percentage of the value for 1990 in each study.

Table 3.1. Estimated dimensions for a typical Western male in 1955 and 2005.

Dimension	1955	2005
height (cm)	171.0	176.5
mass (kg)	72.5	84.0
BMI	24.8	27.0
thigh length (cm)	40.0	44.0
% body fat	20.0	28.0
waist girth (cm)	90.0	93.0

Note: Estimates are based on rates of change from cumulated studies, and current Australian and US datasets.

1.2 Secular changes in the size and shape of the built environment

When humans change shape, the built environment often adapts. Figure 3.2 shows changes in the size of furniture items between the mid-1700s and the late 1900s. The dimensions of chairs, beds, doors, tables and benches have increased to keep pace with changes in body size. This adaptation occurs through a mix of market forces and government and industry regulation. Recently, 'tall' activists successfully lobbied for a 20 cm increase in the height of doors specified by Dutch building codes. In 1997, Brazilian building codes were altered to specify that one-third of all stadium seats should be designed for extra-large people. Sports policy makers regularly discuss resizing infrastructure such as the height of basketball rings and the width of soccer goals.

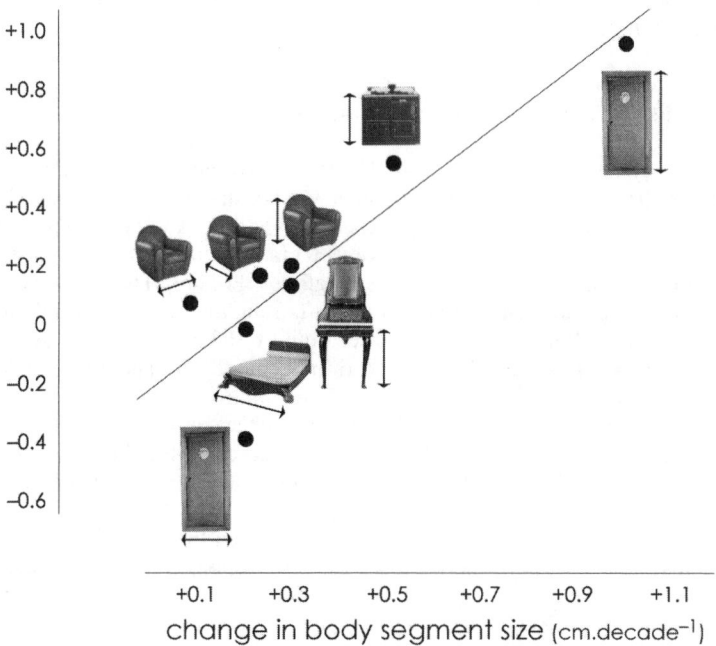

Figure 3.2 Secular trends in the size of furniture items (width and height of doors, height of benches, width of beds, height, depth and width of chairs and tables) between 1750 and 1998, in relation to estimated secular trends in corresponding body dimensions (e.g. stature for door height). Data were collected on items of furniture accessible in Australia (unpublished data).

1.3 Man-machine mismatch

However, this adaptation can be slow, particularly for machines and infrastructure designed to last tens or even hundreds of years. One example is military and civilian aircraft. Because of their enormous development costs, most of the specialized aircraft used around the world are designed to last for 30 years or more. For instance, the F111 has been in service Royal Australian Air Force (RAAF) for 33 years, the Iroquois helicopter for 40 years, and the Caribou for 44 years. The 707 has been flying since 1959. In the US Air Force, the B52 has been flying since 1954.

Where bodies expand to collide with the fixed dimensions of objects, there is a man-environment mismatch which can be resolved only by redesigning environments or by choosing people who can fit into and operate safely in environments. The principle is simple: first match the critical dimensions of the object (height for a doorway, hip girth in a pair of jeans) with the critical dimensions of the body (stature, gluteal girth); then either adjust the object to fit an appropriate proportion of the population (the design solution), or choose the members of the population who can be accommodated by the object (the recruitment solution).

1.4 The military context: recruitment standards

In a military context, the recruitment solution calls for appropriate anthropometric recruitment standards. Military anthropometry recruitment standards have been in place for many years without change. The limits are a standing height of 163 to 193 cm, and a maximum sitting height of 100 cm, maximum buttock-knee length of 63.5 cm and maximum buttock-heel length of 112 cm. These are no longer based on any fit requirements in current aircraft. Standing height is of only indirect relevance in a seated posture, and the lower limit of 163 cm eliminates over 40 per cent of young Australian females. The need to update these standards to accurately reflect fit in the modern Australian Defence Force platforms has become acute.

Both design and recruitment solutions call for accurate anthropometric data. The most recent comprehensive anthropometric survey in the Australian Defence Force dates from 1977, involving 30 measurements on 3000 male personnel (Hendy, 1979) and a survey of aircrew goes back to 1973. Human factors experts in Australia now rely heavily on US data, mainly from the US ANSUR survey in 1987 (Clauser *et al.*, 1988). The need for new anthropometric data was the driver for the establishment of the ADAPT (Australian Defence Anthropometric Personnel Testing) project.

2 METHODS

2.1 Traditional methods

Previous anthropometric surveys with an ergonomics thrust have used physical measurements made with tape measures and calipers, using a 'boundary cases' approach. Usually, about 30 subjects representative of the anthropometric extremes of a population (the 'boundary cases') are assessed for fit and movement in a built environment. Critical environmental dimensions (such as the distance from one seat pan to the back of the next seat in a bus) can then be compared to corresponding body dimensions (such as the buttock to knee length), and judgments made as to what body sizes and shapes can be accommodated. The environment can then be designed to accommodate all but the most extreme bodies.

These methods suffer from a number of limitations. Physical measurements are time-consuming and invasive. The American ANSUR survey, for example, required up to 3–4 hours of measurement time for each subject (Clauser *et al.*, 1988). They also require highly-trained and experienced anthropometrists to be sufficiently accurate. More importantly, only a certain number of dimensions can be captured, and those dimensions might not capture the three-dimensional shape of the body in relevant ways, or in ways which might in future become relevant (for example, with the introduction of a new piece of equipment, or with secular trends in body size). In other words, *a priori* decisions are made about which dimensions are critical. An aircraft designer may be conscious of the need to have sufficient seat-to-seat clearance for leg comfort (though not many seem to have this as a priority), and yet ignore the need for sufficiently wide seats to accommodate large hip breadths. Clearly the choice of 'boundary cases' and critical body and environmental dimensions relies on built-in assumptions. Finally, physical measurement procedures do not allow dynamic modelling – simulating the movements people make as they move about and manipulate their environment.

2.2 Virtual anthropometry

The ADAPT project developed a radically different methodology. It used virtual rather than physical measurements of bodies and environments; it matched every body with each variant of the environment, rather than relying on boundary cases; and it facilitated dynamic task modeling. These advances have been made possible by the recent development of advanced hardware-software suites. The methodology involved three phases:

(1) laser scanning of environments to produce virtual 3D models (Blackwell *et al.*, 2002);
(2) laser scanning of bodies to extract dimensions for rescaling virtual 'manikins'; and
(3) importing the rescaled virtual bodies into the virtual environments and animating them to perform tasks.

2.2.1 *Environment scanning*

Technologies for 3D scanning of objects are well established. For close work, the Faro Arm is typically used (Figure 3.3). A triple jointed arm allows a scanning head to be moved like a paintbrush over the object's surface, creating as it does so a 3D image of the object. Individual movable objects such as hinged doors can be stored as separate files, as can individual points for use as anchor-points. The ADAPT project has scanned 39 crewstations in a range of fast jet, transport, trainer and rotary aircraft. All the scans yield high resolution 3D digital images with a sub-millimetre accuracy. For wider scans (e.g. of a building), scanners such as the Cyrax are often used.

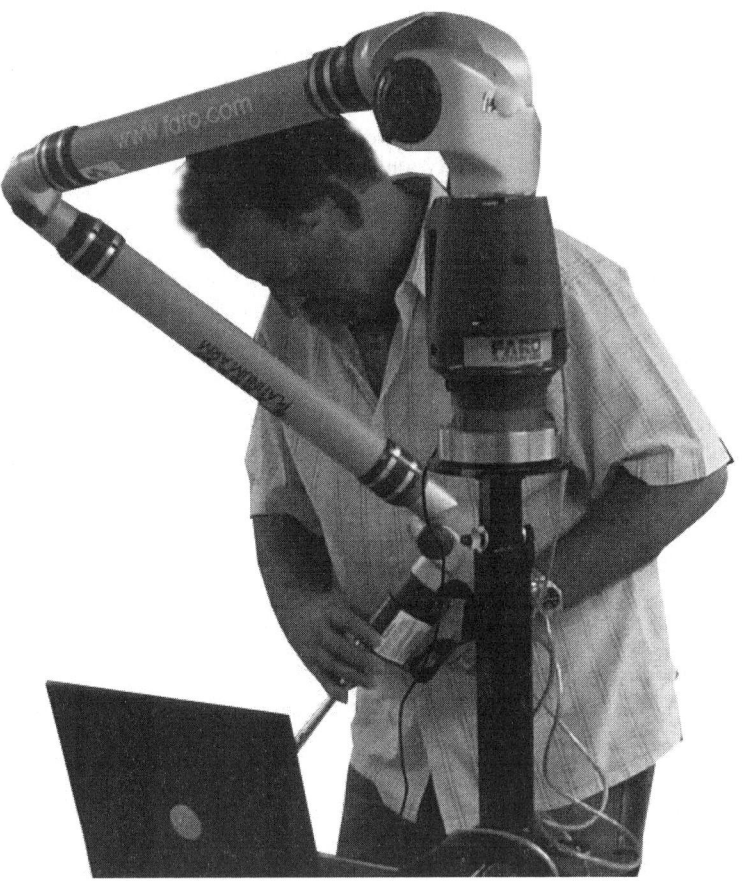

Figure 3.3 The Faro arm scanning device, which is used to scan environments such as crewstations.

2.2.2 3D whole-body scanning

Laser scans of human bodies are less commonly used. Whole-body laser scanners produce 'digital statues' of human bodies. A low-power, eye-safe laser passes down the body in about 10 seconds, and the reflected light is captured by charge-coupled devices linked to video cameras. Computers convert the information into a 'point cloud' of 600,000 points representing the surface of the body (Paquette, 1996). After being cleaned up, the points are joined up into tiny polygonal facets, which can then be smoothed and 'rendered' (the metaphor is from plastering) into a seamless surface (Figure 3.4). The ADAPT project has scanned 1500 young Australians from the general population, with the age (18–30 years) and qualifications typical of the aircrew recruitment population, as well as several hundred serving aircrew.

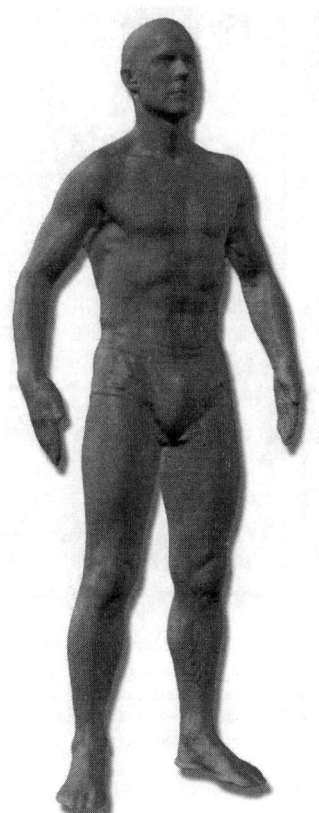

Figure 3.4 A scanned body. A point cloud captured with a laser whole-body scanner has been imported into the figure-posing program *Poser*, and 'rendered' (i.e. given a smooth surface).

While laser body scanning produces an accurate 3D representation of the body, scanned bodies can only be imported into human environments as a rigid and immovable object. It is often very difficult to match scan postures to the target postures in the environment, and this approach also precludes dynamic modeling. Therefore the scan is used primarily to extract measurements which can then be used to rescale animated generic manikins.

Landmarks associated with bony or fleshy prominences on the body can be identified either from physical markers or by automatic software identification. Critical body dimensions are defined in terms of these landmarks. This hardware-software solution generates an extremely high-resolution, reusable representation of the body. Special measurement extraction software can be used to determine point-to-point distances, contour distances across the body surface, cross-sectional and surface areas and volumes (Figure 3.5).

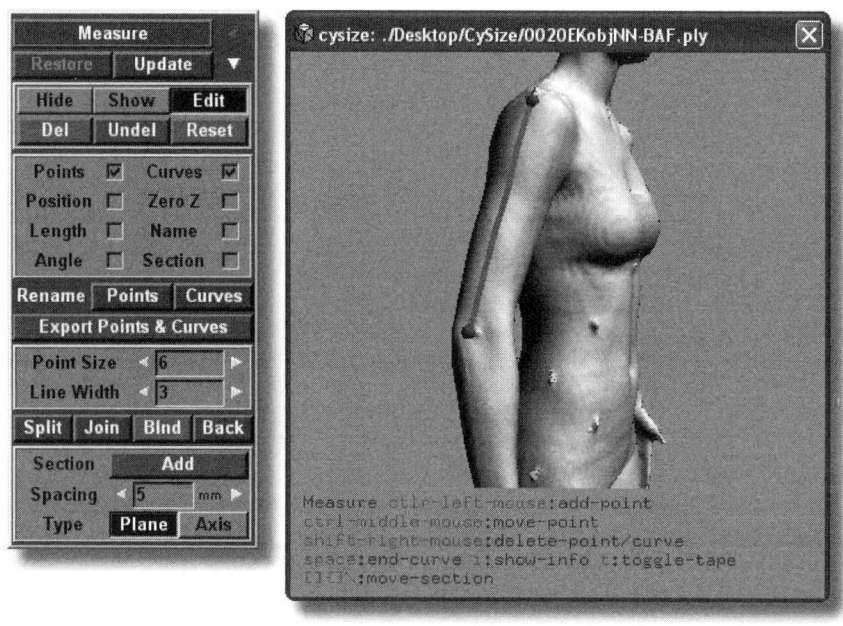

Figure 3.5 An example of measurement extraction software. *CySize* (http://www.headus.com/au/) has the ability to calculate point-to-point measurements, contours and tape measure girths. Small raised triangular prisms used as landmarkers can be seen on the figure. The red line represents Acromiale-Radiale length.

2.2.3 Human modeling programs

In the early 1960s, computer-aided design (CAD) software became available, and aerospace and automotive manufacturers saw the potential for much of the design process to take place in a virtual environment. Recognising the potential to accelerate the design process and at the same time optimize the human-environment interface, digital human modeling tools emerged in the late 1960s in the automotive and aviation industries. With the increasing power of computers, the capabilities of modern human modeling tools have increased dramatically compared to their predecessors developed in the 1960s and 1970s.

Modern day human modeling programs (HMPs), such as *Safeworks*, *Ramsis* and *Jack* use human 'manikins' which have complex kinematic linkages obeying empirically-derived range of motion constraints (Figure 3.6). Onto this rescalable skeleton they overlay a humanoid geometric shell which can be modified according to input data (sex, height, weight). Some HMPs have the ability to reshape the fleshy overlay to match an imported 3D body shape, but the process is not yet automated. Even when an imported 3D shell is used, the geometry of the body merely approximates it by 'growing out' geometrically-defined shapes to meet the 3D envelope. It operates in this way because of the computational demand of translocating hundreds of thousands of datapoints for each manikin. Furthermore, HMPs cannot reproduce the deformation, compression and extension of flesh which occurs as humans move. The HMP selected for the ADAPT project was *Jack*, which was originally developed at the University of Pennsylvania in the 1980s (http://www.ugs.com/products/tecnomatix/docs/fs_tecnomatix_jack.pdf). Using an additional *Jack* software module, the scan of a human can be imported into the *Jack* environment.

Importing clothing and equipment worn by humans also has limitations. The deformation which clothing undergoes during human movement cannot be reproduced in standard programs. Equipment items such as helmets are imported as rigid objects through laser scanning, and 'bolted' onto the body via positioning rules using body landmarks as anchor points. The scan of a standard sized equipment item can be rescaled to simulate different equipment sizes where the different sizes of a given equipment item are isomorphic (e.g. larger and smaller helmets) and matched to individual bodies according to sizing rules (e.g. choosing from available small, medium or large helmet according to virtual head girth).

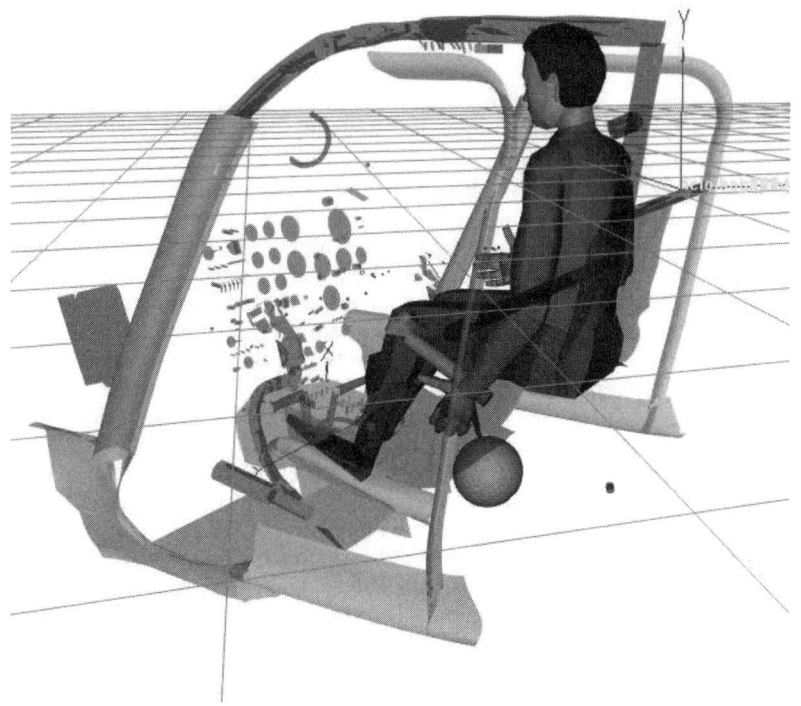

Figure 3.6 *Jack* representation of a rescaled manikin in a helicopter airframe. Note that in *Jack*, various parts of the airframe can be made transparent or switched off (as the instrument panel has been here).

2.2.4 The ADAPT methodology

The overall ADAPT project methodology is shown in Figure 3.7. The left-hand side of the flowchart shows the capture of bodies, and the right-hand side the capture of crewstations. Real bodies are captured by the scanner as point clouds upon which, after conversion to an appropriate file format (in this case .ply), landmarks can be identified and coded as XYZ points. The HMP manikin is then rescaled to these landmark anchor-points. Crewstations are scanned using the Faro arm, and the resulting point cloud is converted to .iges file format, ready to be imported into *Jack*. The rescaled manikins interact with the virtual environment in *Jack*. The HMP runs the manikin through a series of scripted tasks, and issues a report on whether the manikin can fit into the crewstation space, and can perform the scripted tasks. These reports can highlight specific areas of crewstations which cause accommodation problems, and thereby suggest appropriate redesign.

Figure 3.7 The overall project methodology. Human bodies are scanned and captured as point clouds, which are then converted into .ply file format. Measurement extraction programs are used to extract measurements from the 3D representations. These measurements are used to reshape and rescale 'manikins' or generic humans in *Jack*. The manikins then perform certain scripted tasks, such as reaching for controls. Aircraft crewstations are also scanned, captured as point clouds and converted to .iges files. These are imported into *Jack* to form the virtual environment with which the manikins interact. Simulations of bodies, crewstations and task performance can be compared to physical measurements for verification (black shaded boxes).

2.2.5 *Mathematical modeling*

The combination of 3D body scanners, measurement extraction software, object laser scanners and HMPs makes it possible to verify the accommodation and task performance of any individual in any crewstation. However, this process requires a 3D body scan of every individual and processing through the human modeling software. More widespread assessment of potential recruits, such as might occur at the first contact in a 'shop front' environment, requires a set of simplified anthropometric guidelines. To arrive at these guidelines, it is important to isolate a few easily-measured anthropometric dimensions which are strongly predictive of whether an individual will fit in a crewstation or group of crewstations. Discriminant function analysis can be used to determine this set of dimensions. The resultant algorithms act as a 'first filter' during recruitment. Those who pass this stage can then be tested for fit individually in each virtual crewstation. Using these tools, it is also possible to plan an 'anthropometric career path' for any individual. For example, a recruit may not be able to fit into fast jets, but may be suitable for transports or rotary wing aircraft. The mathematical modeling component of the project can also identify 'bottleneck' aircraft in pilot training. These are the aircraft where the largest number of potential recruits fail to be accommodated or to perform the required tasks.

The same strategic approach – general screening algorithms coupled with individual testing in virtual environments – is applicable to a wide range of human-environment interactions. There are obvious military applications in tanks, submarines and other specialist vehicles. There are civilian applications in the workplace, for example on production lines where reach or stature are critical, or where workers are standing close to one another. Equipment, furniture and environments designed for children could be tested by modeling runs on a sample of virtual children.

3 ACCURACY AND VALIDITY

A process as complex as this involves many assumptions. Scanned bodies and crewstations are represented as rigid objects, whereas in reality flesh, seats and equipment are deformable and compressible. Human joints are much more complex than the linkage models used in even the most advanced HMPs, and people use many different movement strategies to accomplish the same task. There is considerable inter- and intra-subject variability in joint ranges of motion. What confidence can we have that the representations of bodies and environments in HMPs match reality?

Measurements derived from scans and from HMP representations can be compared with physical measurements at a number of points. These comparisons can be used to quantify the accuracy of the models. The ADAPT project has extensively tested the accuracy of each step in the pathway shown in Figure 3.7, including the measurement of task performance in real crewstations (Figure 3.8).

The determined accuracy will be specific to the software-hardware suites used, but comparisons using alternative body capture and measurement extraction software have yielded broadly similar results. When compared to physical measurements taken by expert anthropometrists, measurements extracted from 3D scans differ by an average of 5 mm (95 per cent confidence limits –9 to +20 mm). The dimensions of the rescaled manikin in the HMP differ by an average of 7 mm (95 per cent confidence limits –16 to +31 mm), and task simulation is accurate to 5 mm on average (95 per cent confidence limits –35 to +46 mm). These data suggest that for most ergonomic modeling the accuracy of virtual anthropometry is satisfactory.

Figure 3.8 Testing of a subject in the Kiowa helicopter as part of validation trials. Here physical reach is being measured, which is later compared to the simulated value for the manikin in *Jack*.

4 CONCLUSION

'Virtual anthropometry' has several limitations. It requires very expensive equipment and highly-trained operators. It still depends on physical landmarking, and loses some accuracy in making digital measurements. Scanners also cannot measure skinfolds. Nevertheless, virtual anthropometry expands the horizons of traditional anthropometry through its ability to measure surface and cross-sectional areas and volumes, and its capacity to model dynamic interactions. The new anthropometry has been made possible by the recent rapid expansion of the power of computing, which it exploits to match large numbers of bodies against many environments on very large numbers of variables, thus avoiding the presuppositions and approximations of the 'boundary case' approach. It is sufficiently accurate for quite precise tasks such as matching pilots to military aircraft.

5 ACKNOWLEDGEMENTS

The authors would like to acknowledge the financial support of the Australian Defence Force through Project MIS872, and the contribution of the ADAPT Consortium partners: The University of South Australia, Sinclair Knight Merz, Permian, the University of Ballarat and the Australian Institute of Sport.

6 REFERENCES

Blackwell, S., Robinette, K.M., Boehmer, K., Fleming, S., Kelly, S., Brill, T., Hoeferlin, D., Burnsides, D. and Daanen, H., 2002, *Civilian American and European Surface Anthropometry Resource (CAESAR), Final Report, Volume II: Descriptions.* Warrendale, PA: AFRL-HE-WP-TR-2002-0173, Air Force Research Laboratory, Human Effectiveness Directorate, Crew System Interface Division, 2255 H Street, Wright-Patterson AFB OH 45433-7022 and Society of Automotive Engineers International, 400 Commonwealth Drive, United States Air Force Research Laboratory.

Clauser, C., Tebbets, I., Bradtmiller, B., McConville, J. and Gordon, C.C., 1988, *Measurer's Handbook: U.S. Army Anthropometric Survey 1987–1988.* Natick, MA: United States Army Natick Research, Development and Engineering Center.

Hendy, K.C., 1979, *Australian Tri-Service Anthropometric Survey, 1977: Part 2: Survey Results: Combined Services Aircrew Group,* Melbourne: Aeronautical Research Laboratory.

Meredith, H.V., 1976, 'Findings from Asia, Europe and North America on secular change in mean height of children, youth and young adults', *American Journal of Physical Anthropology,* **44**, 315–326.

Paquette, S., 1996, '3D scanning in apparel design and human engineering', *IEEE Computer Graphics and Applications,* **16(5)**, 11–15.

CHAPTER FOUR

A comparison of the accuracy of the Vitus Smart® and Hamamatsu Body Line® 3D whole-body scanners

N. Daniell

Centre for Applied Anthropometry, University of South Australia

1 INTRODUCTION

Prior to three-dimensional (3D) scanning, anthropometrists were required to manually measure the human body using traditional tools like anthropometers and tape measures. In the last decade, 3D scanning has progressed rapidly with the accuracy and reliability now acceptable for many interested industries such as clothing, textiles, sports science, ergonomics and anthropometry (Istook and Hwang, 2001).

The technology of 3D scanning will be explained in this paper along with how it can be applied to anthropometry. This paper will focus on a study that determined the accuracy of the Vitus Smart® and Hamamatsu Body Line® 3D whole-body scanners. The study compared the difference between physical body measurements and measurements derived from the hardware-software suites used by the Vitus Smart and Hamamatsu scanners. The components of error at each stage of the process were also quantified. Results indicate that 3D scanning is a viable and accurate option for anthropometry in the future.

2 HOW 3D SCANNERS WORK

3D scanning is a new technology that allows the outer body of a human to be represented as a 3D object. Development of this technology required experts from various areas including optics, image processing and mechanics. Daanen and Van de Water (1998) stated that the following characteristics remain consistent in all 3D scanners:

- The scanner requires a light source of some form to be projected onto the body.

- This light source has to be captured by cameras when it meets the body surface.
- Compatible software is required to extract the 3D structure of the object's surface.
- Finally a computer system is necessary to visualise the data captured by the scanner.

2.1 Light source

Although these features remain constant, whole-body scanners vary significantly between different manufacturers and machine models. The most significant variation in 3D scanners is the light source used to capture the human body as a 3D image. Currently there are two light sources in use, laser and white light.

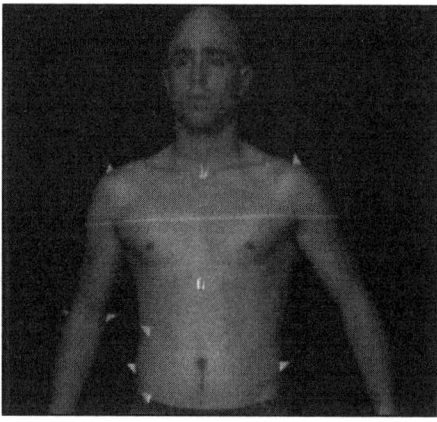

Figure 4.1 Laser projected onto the human body as a thin and sharply-defined stripe.

A laser scanner consists of a laser, an optical system and a light sensor. The scanner projects lasers onto the human body as a thin and sharply-defined stripe (Figure 4.1). Lasers and cameras are positioned at various angles to ensure the entire body is captured without the subject having to be repositioned. The lasers are aligned horizontally and move vertically along the body to capture the entire surface. Most laser scanners take ~10 s to scan the entire body. The light sensors (cameras) capture the laser's reflection off the body as a series of points, often referred to as "point cloud" data. Normally a scan will contain 500,000–700,000 data points, each with a XYZ (X, medio-lateral; Y, anterio-posterior; Z, vertical) coordinate.

Laser-based scanners are considered by many as the most accurate light source, but they are also the most expensive (Daanen and Van de Water, 1998). One major benefit of some laser scanners is their ability to capture colour or luminance. This is not possible with any other light source. A cheaper alternative to laser scanners is white-light scanners.

White-light projection does not require moving parts; instead a light pattern is projected onto the body, normally a series of horizontal white stripes (Figure 4.2). Similar to laser scanners, the reflection of the light off the body is captured by cameras surrounding the body and calculated as XYZ coordinates. Several scanning companies using white light, including TC2, have employed a technique known as Phase Measuring Profilometry (PMP) (Paquette, 1996). This method enhances the scan quality by moving the horizontal white stripes at set intervals.

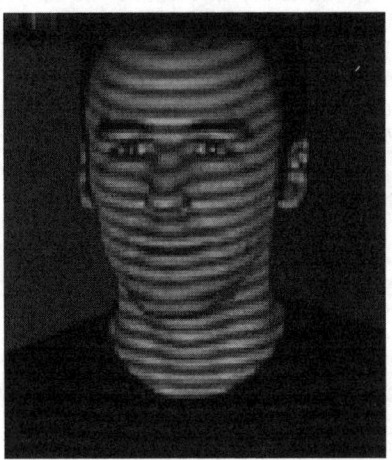

Figure 4.2 White light scanner projecting a series of horizontal white stripes across the subject (InSpeck, 2004).

The five leading 3D scanning companies are Cyberware (Monterey, USA), Human Solutions/Vitronic (Wiesbaden, Germany), Hamamatsu (Hamamatsu City, Japan), Hamano (Kanagawa, Japan) and TC2 (Cary, USA). The scanners used in the present study were from Human Solutions/Vitronic and Hamamatsu.

2.2 Landmark identification

Most scanners have associated measurement extraction software that allows the scans to be uploaded and analysed. They use varied landmarking systems with most software programs being designed around the requirements of a specific 3D scanning company. All programs used by the leading 3D scanning companies are able to take point-to-point measurements and girths. Some of the programs are also able to measure contour measurements, cross-sectional areas, volumes and surface areas.

To accurately record measurements, a reliable method is required for identifying landmarks on a scanned body. Traditional anthropometry has focused on using bony points as defined landmarks; these cannot always be visually located on a 3D scan. As a result, differing landmarking methods have been developed for 3D scanning. There are currently three recognised ways to identify landmarks in the associated software programs.

Automatic Landmark Recognition (ALR) is the most popular method with no manual identification required. The software programs use algorithms to locate critical landmarks and calculate the required measurements (Mckinnon and Istook, 2002). This is an extremely fast process capable of producing over 100 different measurements in ~30 s. The algorithms can only use the exterior surface to determine the landmark's position, therefore assuming standard algorithms for different somatotypes. The error using this method is unacceptable for anthropometry due to the difficulty in locating the landmarks correctly without palpation (Burnsides, Boehmer and Robinette, 2001).

Digital Landmark Recognition (DLR) requires the landmarks to be manually located by a trained observer on the scanned image in the software programs. DLR can produce accurate and precise results for clearly identifiable landmarks such as the Supramenton, Meta carpals and Dactylion. However, the method is unacceptable for most landmarks due to the location being unidentifiable without palpating the exterior surface of the body.

Physical Digital Landmarking (PDL) is a lengthy process requiring physical landmarkers to be positioned on the body, making the experience more uncomfortable for the subject (Certain and Stuetzle, 1999). The landmarkers are then manually identified as above, using a set protocol. Manually identifying the landmarkers is not required for some scanners if the software is able to detect luminance or colour stickers that have been placed on the body. PDL is considered the most accurate method and is the chosen method for this study.

3 METHODS

3.1 Participants in study

The participants for this study were 17 males and 13 females aged 18–25 years. The participants' characteristics are shown in Table 4.1. To be eligible for this study, potential participants must not have had a serious injury (e.g. broken bone) that had affected their gross morphology.

Table 4.1 Age (y), mass (kg), stature (cm) and Body Mass Index (BMI, $kg.m^{-2}$) of the participants.

Sex	N		Age (y)	Mass (kg)	Stature (cm)	BMI ($kg.m^{-2}$)
Male	17	$\bar{x}$	20.6	73.4	179.4	22.9
		SD	1.9	8.8	9.0	2.6
Female	13	$\bar{x}$	20.3	60.7	169.1	21.2
		SD	0.95	7.3	3.8	2.4

3.2 Measurements used

All physical measurements were taken following the protocols described by the International Society for the Advancement of Kinanthropometry (ISAK, 2001). The physical measurements were taken by an anthropometrist accredited at Level 2 by ISAK.

Thirteen duplicate anthropometric measurements were taken in this study (height and mass were also recorded). These measurements included five girths, four lengths and four breadths, all taken according to ISAK protocols.

After being physically measured, participants were scanned using *Vitus Smart* and *Hamamatsu Body Line* scanners. All participants adopted the Standard Scanning Pose (SSP) prior to scanning (see Figure 4.3). The SSP requires the participant to be positioned with the:

- head in the Frankfort plane (Norton *et al.*, 2004); that is, the participant looks straight ahead and the Tragion is aligned horizontally with the Infraorbitale;
- shoulders relaxed with arms slightly abducted and elbows slightly flexed;
- palms of each hand facing the body, with fingers extended and together, and the thumb pointing forward;
- feet approximately shoulder width apart.

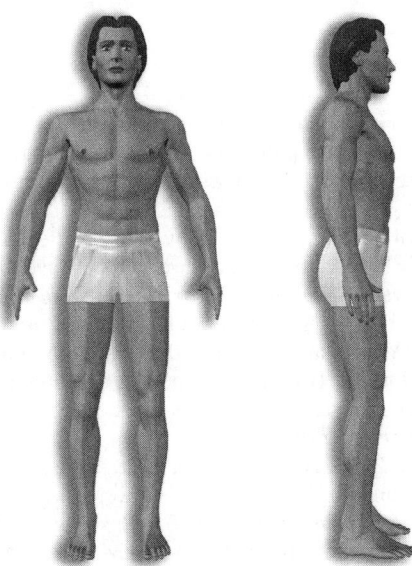

Figure 4.3 Standard Scanning Pose (SSP) that the participants were required to adopt during all scans for the study.

The SSP is adopted to minimise missing data (e.g. gaps under the arms) and to help facilitate measurements. Participants were instructed to breathe naturally during the scanning process and to keep their muscles relaxed. All participants wore form-fitting underwear. Dark-coloured and reflective materials could not be worn as they are poorly captured by the scanner. A swim cap was also worn to help capture the head and to minimise the effect of hair on head girth.

All participants were landmarked for scanning using the same sites previously marked for the physical measurements. A Kupke Triangular Prism (KTP) physical landmarker was applied at each of the sites using double-sided adhesive tape. The KTP (Figure 4.4) pointed distally to the marked landmark for more accurate identification in the measurement extraction software. Luminance stickers were applied with the centre of the sticker placed onto the landmark.

Figure 4.4 A custom-made KTP landmarker. The landmarker is attached so that it points down towards the landmark.

3.3 Data flow

Figure 4.5 shows a flow chart illustrating the data flows in this study. This flow chart shows three predominant pathways after the landmarked bodies. These are traditional anthropometry (left), the *Vitus Pathway* (centre) and the *Hamamatsu Pathway* (right).

The first stage of the *Vitus Pathway* was to scan each participant using the *Vitus Smart* scanner. The captured data were then saved as a point cloud by Vitronic's native software *ScanWorx Editor*. There were a number of file format conversions required before the data could be used in downstream software applications. The .PLY format represents data as a polygonal mesh, and is the required file format for the two measurement extraction programs used in this pathway, *DigiSize* and *CySize*. *Digisize* was used to manually extract the lengths and breadths, and *CySize* the girths.

The *Hamamatsu Pathway* is similar to the Vitus. Following landmarking, each participant was scanned. The captured data were displayed as a point cloud by Hamamatsu's native *Body Line Manager* (*BLM*) software. The point cloud data were converted to a polygonal mesh by the *BLM* software in proprietary .ABL format. All measurements were then extracted using the *BLM* software.

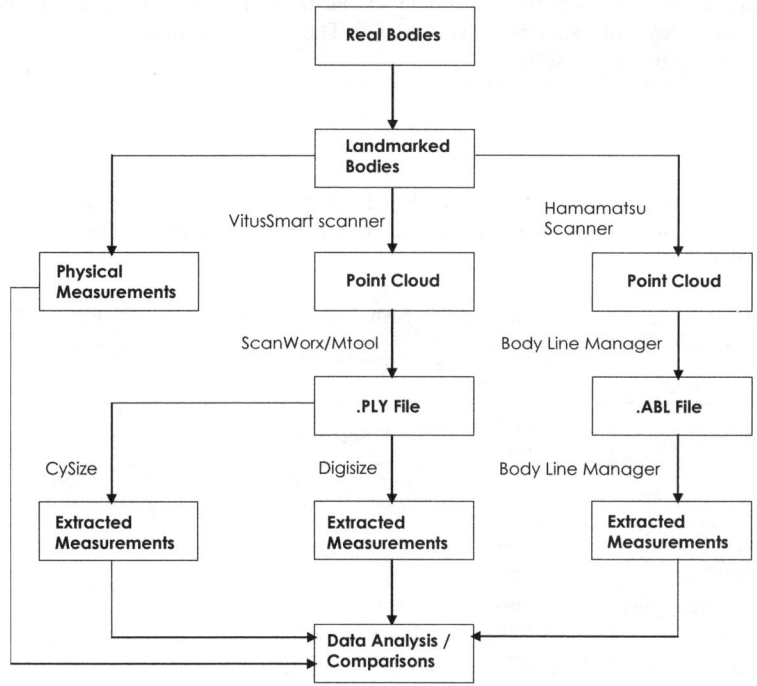

Figure 4.5 Data flow chart for this study: .PLY and .ABL are proprietary file formats,
PDL = physical digital landmark, ALR = automatic landmark recognition.

3.4 Statistics used

Bland-Altman analysis was used in this study. This technique allows identification
of systematic error (bias) and random error (standard error of measurement; SEM),
and uses mean-difference plots as the main graphical display method (Atkinson and
Nevill, 1998). Bland-Altman analysis provides a range within which 95% of future
calculated differences should fall.

In this report, bias is quantified as the physical measurement minus the scan-
derived measurement and is reported in mm. Therefore, a positive bias indicates a
larger physical measurement and a negative bias a larger scan measurement.

The statistic used to quantify random error is the SEM. The SEM reflects how
closely the differences between measurements are clustered together. The equation
for the SEM is:

$$\text{SEM} = \text{SD}_{\text{diff}} / \sqrt{2} \quad (4.1)$$

where SD_{diff} is the standard deviation of the difference scores (mm).

Mean biases between the *Vitus* and *Hamamatsu Pathway* were compared using paired t-tests, with an alpha level of 0.05. The same type of paired t-test was used to compare the mean SEMs.

4 RESULTS

Table 4.2 summarises the bias and SEM associated with each of the 13 measurements for each scanner. Also shown are the mean values for physical measurements.

Table 4.2 Mean values for each of the physical measurements, and biases and standard errors of measurement (SEMs) for comparisons between physical and scan-derived measurements. NA = not available.

Measurement	Mean (mm)	Hamamatsu		Vitus Smart	
		Bias (mm)	SEM (mm)	Bias (mm)	SEM (mm)
Acromiale-Radiale	324	11.6	4.3	4.4	8.1
Radiale-stylion	250	−7.3	4.5	−9.6	5.7
Midstylion-dactylion	194	7.6	3.8	3.8	3.5
Biacromial	381	12.8	5.3	3.2	6.0
Biiliocristal	273	−25.1	9.8	−28.0	9.1
Foot Length	256	−1.7	2.3	−0.8	3.8
Biepicondylar humerus	69	−4.2	3.3	1.5	3.0
Biepicondylar femur	96	−5.5	4.2	−17.0	6.1
Head girth	564	−18.4	8.3	NA	NA
Arm girth	293	−1.6	3.7	−7.0	4.9
Waist girth	754	−7.6	8.1	−7.1	11.2
Calf girth	369	−4.1	2.2	−1.1	3.8
Ankle girth	220	−11.8	2.5	−0.9	4.4
ALL SITES		−4.3	4.8	−5.1	5.8

The mean biases for comparable measurements (i.e. excluding head girth) were −4.3 mm for the *Hamamatsu Pathway* and −5.1 mm for the *Vitus Pathway*. These were not significantly different when compared using a paired t-test. The mean SEMs were 4.8 mm for the Hamamatsu and 5.8 mm for the Vitus Smart. Despite a difference of only 1.0 mm, the mean SEM was significantly smaller for the Hamamatsu (p = 0.01).

4.1 Example of Bland-Altman analysis

Figure 4.6 is an example of how Bland-Altman plots were used for analysis in this study. The example given is for Acromiale-radiale comparisons. The results for the Acromiale-radiale illustrate greater mean bias in the Hamamatsu, but greater random error in the Vitus scanner.

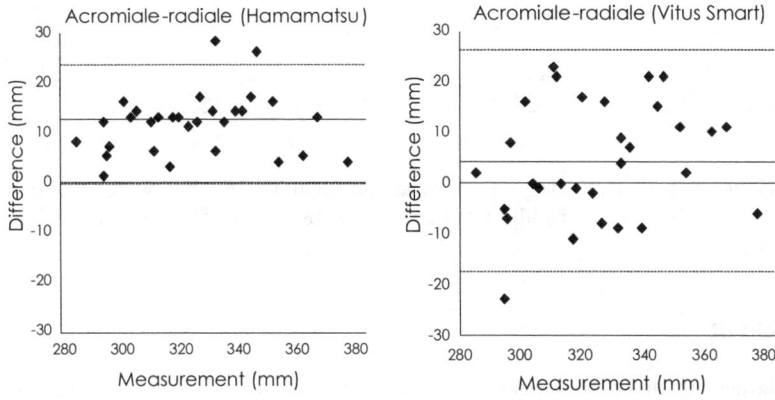

Figure 4.6 Bland-Altman plots of the physical measurements compared to Hamamatsu measurements (left panel) and the Vitus Smart (right panel) for Acromiale-radiale length.

4.2 %SEMs

Table 4.3 SEMs of each scanner for lengths and breadths, also presented as a percentage of the mean physical measurement (%SEM). SEMs that are underlined have a difference of >2 % compared to the mean measurement.

Measurement	Hamamatsu		Vitus	
	SEM	%SEM	SEM	%SEM
Acromiale-radiale	4.3	1.3	8.1	<u>2.5</u>
Radiale-stylion	4.5	1.8	5.7	<u>2.3</u>
Midstylion-dactylion	4.9	<u>2.5</u>	4.0	<u>2.1</u>
Biacromial	5.3	1.4	6.0	1.6
Biiliocristal	9.8	<u>3.6</u>	9.1	<u>3.3</u>
Foot Length	2.3	0.9	3.8	1.5
Biepicondylar humerus	3.3	<u>4.8</u>	3.0	<u>4.3</u>
Biepicondylar femur	4.2	<u>4.4</u>	6.1	<u>6.4</u>
ALL SITES	4.8	2.6	5.7	3.0

Table 4.4 SEMs of each scanner for girths, also presented as a percentage of the mean measurement. No SEMs had a difference of over 2% compared to the mean measurement. NA = not available.

Measurement	Hamamatsu		Vitus	
	SEM	%SEM	SEM	%SEM
Head girth	8.3	1.5	NA	NA
Arm girth	3.7	1.3	4.9	1.7
Waist girth	8.1	1.1	11.2	1.5
Calf girth	2.2	0.6	3.8	1.0
Ankle girth	2.5	1.1	3.8	1.0
ALL SITES	5.0	1.4	6.0	1.3

SEMs were typically 2.0–2.5% of mean physical values, but were significantly larger for Biiliocristal breadth (3.5%), Humerus breadth (4.5%) and Femur breadth (5.4%).

5 ERROR

One aspect of this study aimed to quantify the level of error at each stage of the *Vitus* and *Hamamatsu Pathway*. Figure 4.7 shows the random error for each of the stages within both hardware-software suites. The errors are also reported in Table 4.5. All errors calculated for this process were expressed as SEMs. Once the error at each stage was determined, the total random error (TRE) was calculated. Since the errors at each phase are independent of one another, the total random error is the square root of the sum of all the individual errors squared.

$$\text{TRE} = \sqrt{\sum (e_1)^2 + (e_2)^2 + ... + (e_n)^2} \quad (4.2)$$

All are expressed as SEMs except the Bias value. The bias is the precision of each scanner. The total error has accounted for the four components of error that were quantified in this study. The whole-of-process error is the average SEM for physical measurements vs. scan-derived measurements.

Table 4.5 A summary of the errors quantified in this study.

Error	Hamamatsu	Vitus
Physical landmark location	2.9	2.9
Point cloud data (polygon)	0.9	1.1
Registration (calibration) error	1.7	1.7
Software landmark location	0.15–3.2	3.2–3.6
TOTAL ERROR	3.5–4.7	4.8–5.0
Whole of process error (SEM)	4.8	5.7–6.0
Whole of process error (Bias)	–4.3	–5.1

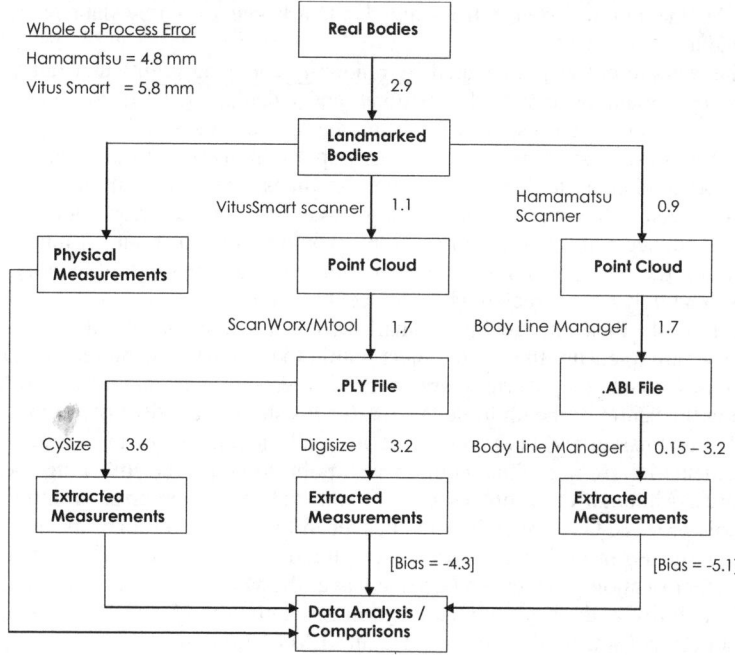

Figure 4.7 Data flow chart for the study with the errors associated with each stage of the pathways. The errors are all represented as SEMs (mm) unless otherwise stated.

5.1 Error in physical landmark location

The level of error that occurs during physical landmark location will be consistent between all three pathways; the physical measurements, the *Vitus Pathway* and the *Hamamatsu Pathway*. The size of this error is dependant on the anthropometrist's ability to locate the correct landmarks on each subject. The completion of a reliability study by University of South Australia found the results established an average inter-tester SEM of 2.9 mm for the physical location of landmarks by two experienced anthropometrists. The SEM of 2.9 mm will be accounted for in both the *Vitus* and *Hamamatsu Pathways*.

5.2 Error in *Vitus Pathway*

The first step in the *Vitus pathway* is the scanning of the subject. The 3D scan is uploaded onto the computer as a series of raw data points. This automatically introduces a small error. There are very small gaps between the raw data points due to the scanner's inability to capture the entire body. When the data points form a polygon mesh they are joined adjacently forming small triangles between the points. Landmarks are likely to fall within these triangles and not directly on a data point. The maximum distance from any landmark site to a raw data point was calculated at ~1.6 mm.

The random error is calculated by randomly choosing points that fall within the triangle, simulating landmark positions, and calculating the distance between the landmark and the closest data point, an average distance of 0.8 mm (SEM of 0.28 mm) was established. However, an assumption is made that the surface of the body is perpendicular to the laser stripe. At times, the laser will hit the body obliquely, causing the triangles to be larger than those as described above. With this taken into account the SEM for this error would double to about 1.1 mm.

During the file conversions, the point cloud data are meshed into a polygonal network. The polygon is created when adjacent raw data points are joined to form a triangular mesh, thus introducing a small error. The raw data points from each camera overlap to ensure the entire object within the scanner is captured. When the polygon is formed, the overlaps are merged, therefore introducing an error by predicting the centre of the multiple layers. Results showed an SEM of 1.7 mm.

The landmarkers (KTPs) have to be manually identified for the extraction of measurements in *Digisize*. This requires each point to be picked following specific procedures. Although these procedures are well defined, some error is introduced during this process. Our research group has recently conducted reliability studies to determine the error during measurement extraction. Researchers found that the manual identification process for *Digisize* has an SEM of 3.2 mm for this study. A similar reliability study was conducted by our research group for *CySize*, although a different protocol was used. The error had an SEM of 3.6 mm.

To determine the total error of the *Vitus Pathway*, error in each stage must be accounted for. This will be completed separately for the lengths/breadths and then again for the girths due to the different errors associated with software landmark identification. The TRE equation was used (4.2).

The total error in the *Vitus Pathway* was 4.8 mm for lengths and breadths with the error in girths being slightly higher at 5.0 mm.

The SEMs varied between landmark sites with the averages used for this equation.

Results show that the whole-of-process SEM was slightly larger for girths (6.0 mm) than for lengths and breadths (5.7 mm). The SEMs varied between landmark sites.

The whole-of-process error for the *Vitus Pathway* was calculated by comparing the physical measurements to the *Vitus Smart* scan-derived measurements. The measured whole-of-process error (SEM = 5.7 to 6.0 mm) was very close to the total random error (SEM = 4.8 to 5.0 mm), especially given that some error was unaccounted for. The error associated with respiration and postural sway was not quantified and is likely to have contributed to this difference. Waist girth (min) did show some variation with the largest SEM (11.2 mm) of all measurements taken, although it still had a %SEM of <2%. It is likely that this error was partially due to ventilation.

5.3 Error in *Hamamatsu Pathway*

The error involved with the *Hamamatsu Pathway* is quite similar to that of the Vitus. Again the first type of error is due to the gaps between the raw data points. The SEM was calculated using the same process as the Vitus with a resultant error of 0.9 mm. The error was slightly less due to a higher resolution with the scanner.

The overlapping data also produce some error with the Hamamatsu scanner when a polygonal mesh is created. Due to BLM removing the overlaps automatically, the assumption is made that the error is similar to the *Vitus Pathway*, with an SEM of 1.7 mm.

The last type of error in the *Hamamatsu Pathway* is the error in locating landmarks through *BLM*. The landmarkers used for this pathway were KTPs and luminance stickers. KTPs were only used at landmark sites where the stickers were not effectively captured by the scanner. The error in identifying KTPs is identical to the *Vitus Pathway*, as the same procedures in identification were used. Therefore the SEM is 3.2 mm.

The luminance stickers are automatically identified by the software and require no manual identification. The distance between the predicted landmark position and the real centre of the virtual sticker was calculated using valid statistical techniques. The mean difference was calculated at 0.24 mm, and the SEM was 0.07 mm. However if the sticker is oblique to the laser then the error would increase as fewer points would fall within the luminance sticker. The SEM is still likely to be <0.15 mm. This would almost eliminate error in landmark identification providing the stickers were reliable.

The total error for the *Hamamatsu Pathway* was calculated using the same procedure as the *Vitus Pathway* (Equation 4.2). When using KTPs as landmarkers the total error was calculated at 4.8 mm.

If only luminance stickers were used in the study then the 3.2 mm error for software landmark location would be significantly reduced to <0.15 mm. This would reduce the total error to 3.4 mm. However, it cannot be concluded that the total error is 3.4 mm for the *Hamamatsu Pathway* due to the inconsistency of the software identifying the location of the luminance stickers.

The whole-of-process error for the *Hamamatsu Pathway* was calculated by comparing the physical measurements to the Hamamatsu scan-derived measurements. The mean SEM for the 13 measurements was 4.8 mm. The estimated overall error (by summing components of random error) was 3.5 to 4.7 mm, this was dependent on the landmarkers used. Although the upper limit of the total error is close to the whole-of-process error, it would be slightly less as some landmarkers were luminance stickers. Once again, the measured whole-of-process error was very close to the sum of the component errors. Like the Vitus Smart, the error associated with respiration and postural sway may have contributed to the differences in error. Overall, the errors were smaller for the *Hamamatsu Pathway* when compared to the *Vitus Pathway*.

6 DISCUSSION

6.1 Measurement artefacts

Biases between physical and scan-derived measurements may to some extent be explained by compression. The measurement of bony breadths with large and small sliding calipers calls for pressure to be applied at the bony landmarks, and girth tapes cause small indentations of the skin, whereas the scan interprets the body as rigid and incompressible. This would partly explain why scan-derived measurements were 25–28 mm larger for the biiliocristal breadth, 6–17 mm larger for the biepicondylar femur, and 2–4 mm larger for the biepicondylar humerus. Going against this trend was biacromial breadth where the scanned measurement was unexpectedly smaller than the physical measurement. This may be due to the acromion sites moving as the arms are slightly abducted in the standard scanning pose. Scan-derived girths were also marginally greater (1–12 mm).

A second artefact which might mitigate biases is the way in which the Hamamatsu software calculates girths. While a "slanting line" option allows the user to measure girths perpendicular to the long axis of the arm, this technique is currently not available for the legs. All lower limb girths are taken parallel to the standing surface. Because the subject is scanned with feet shoulder width apart, the plane used to calculate calf and ankle girth for the Hamamatsu scan cuts the body slightly obliquely, producing artificially large girths.

This difference only accounts for a small percentage of the differences found between physical measurements and Hamamatsu lower limb (calf girth and ankle girth) measurements. The bias for the measurements was –4 mm and –12 mm respectively, indicating that the Hamamatsu consistently overestimates lower leg measurements. The Hamamatsu Corporation is currently working on a "slanted curve option" for the legs.

Waist girth was physically measured at the end of a normal expiration (end tidal), while the participants were asked to remain still during the scan, not necessarily at end-tidal. This would tend to inflate the random error, as participants may have been at different stages of a shallow breathing cycle.

Given these considerations, scan-derived measurements show relatively small biases and random errors relative to physical measurements.

6.2 Limitations

During this study a number of limitations were established regarding scans and scan-derived measurements. The first relates to landmarker identification. The Hamamatsu was inconsistent in its ability to identify the luminance stickers. Trials were conducted to check whether the luminance stickers would be identified at each landmark site. On average only about two thirds of the luminance stickers were effectively identified at the required landmark sites. The *Vitus Smart* is limited by its inability to use luminance. The KTPs are ideal for the Vitus as they were produced by our research team specifically for use in the *Vitus Smart* scanner. If the luminance stickers become more reliable they will be extremely effective landmarkers and substantially reduce errors involved with 3D scanning.

A second problem relates to the non-convertibility of proprietary file formats. Vitronic's .CSF and Hamamatsu's .ABL formats are not convertible to other formats using third-party file format translators. In particular, Hamamatsu's raw data cannot be exported at all and so cannot be used in third-party measurement extraction software, such as *DigiSize*. This issue led to the decision to compare two hardware-software suites instead of attempting to simply compare scanners.

Missing data are a problem with the *Vitus Smart* and *Hamamatsu* scanners. The scanners cannot pick up shadowed areas such as under the arms or the crotch, and also perform poorly on areas where the laser strikes the body at a very oblique angle, such as the top of the head and the shoulders. Missing data on the inside of the upper arm also impaired the ability to take arm girths. The Hamamatsu *BLM* software has an auto-fill and smooth function, and *CySize* had a custom-programmed auto-fill function. However, the accuracy of these functions remains to be assessed.

7 CONCLUSION

The data presented here show that both the *Vitus* and *Hamamatsu Pathways* have very similar accuracies when compared to the physical measures, which is not surprising given that the scanners use similar technologies and the analysis software suites use similar algorithms. The biases and random errors are sufficiently small to suggest that 3D anthropometry may offer a viable and richer alternative to traditional methods.

8 REFERENCES

Atkinson, G. and Nevill, A.M., 1998, Statistical methods for assessing measurement error (reliability) in variables relevant to sports medicine. *Sports Medicine,* **26**, 217–238.

Burnsides, D.B., Boehmer, M. and Robinette, K.M., 2001, 3-D landmark detection and identification in the CAESAR project. In *Proceedings of the Third International Conference on 3-D Digital Imaging and Modelling Conference.* (Quebec City: IEEE Computer Society).

Certain, A. and Stuetzle, W., 1999, Automatic Body Measurement for Mass Customization of Garments. In *Proceedings of the Second International Conference on 3-D Imaging and Modelling Conference* (Ottawa: IEEE Computer Society), pp. 405–414.

Daanen, H.A.M. and Van de Water, G.J., 1998, Whole body scanners. *Displays,* **19**, 111–120.

InSpeck, 2004, InSpeck: 3D Gallery, Retrieved June 10, 2004, from website: http://www.inspeck.com/.

International Society for the Advancement of Kinanthropometry (ISAK), 2001. *International Standards for Anthropometric Assessment.* (Underdale, SA, Australia.)

Istook, C.L. and Hwang, S., 2001, 3D body scanning systems with application to the apparel industry. *Journal of Fashion Marketing and Management,* **5**, 120–132.

Mckinnon, L. and Istook, C., 2002, Body scanning: The effects of subject respiration and foot positioning on the data integrity of scanned measurements. *Journal of Fashion Marketing and Management,* **6**, 103–121.

Norton, K., Wittingham, N., Carter, L., Kerr, D., Gore, C. and Marfell-Jones, M., 2004, Measurement Techniques in Anthropometry. In *Anthropometrica,* edited by Norton, K. and Olds, T. (Sydney: UNSW Press), pp. 27–75.

Paquette, S., 1996, 3D scanning in apparel design and human engineering. *IEEE Computer Graphics and Applications,* **16**, 11–15.

CHAPTER FIVE

An anthropometric method of measuring standing posture with 3D analysis

G.R. Tomkinson and L.G. Shaw

School of Health Sciences, University of South Australia

While standing posture has traditionally been visually observed, complementary devices capable of directly measuring posture are gaining in popularity. Recent developments in three-dimensional (3D) technology provide new and exciting opportunities for postural measurement. With the availability of a plethora of measurement techniques, and without general agreement on how to measure posture, there exists a need for a universally-accepted method of measuring standing posture. The aim of this study was to describe an anthropometric method of directly measuring standing posture with 3D analysis.

To do this, a systematic strategy which identified, defined, described, operationalized and calculated standing postural measurements, was adopted. An extensive review of literature was initially conducted to identify studies which explicitly reported on standing postural measurement. Identified measurements were defined and described, with all descriptions operationalized as relationships between landmarks. These landmarks were then used as inputs into calculations to quantify the postural measurements.

Not only can this anthropometric method of directly measuring standing posture be used with 3D analysis, but it can also be adapted for use with other measurement techniques.

1 BACKGROUND

Posture is generally defined as the relative arrangement of the parts of the body. Historically, it has been visually observed by experienced clinicians (e.g. physiotherapists, chiropractors, orthopedic surgeons, ergonomists) rather than directly measured, with the aim of determining whether the observed posture contributes to the presenting condition(s) (Arnold et al., 2000). This is despite a lack of supportive evidence for a link between posture and musculoskeletal pain, or posture and other clinical symptoms (e.g. loss of function) (Arnold et al., 2000; Raine and Twomey, 1994; Tyson, 2003). It has often been reported that visual observations of posture exhibit poor agreement between repeated measurements (e.g. Fedorak *et al.*, 2003), which reduces the confidence in single measurements and the ability to track changes over time. Factors affecting the repeatability of visual observations include the lack of use of standardized measurement protocols, ambiguous classification criteria, the inconsistent use of standardized postures, and the unwillingness to use "reference" landmarks.

Postural measurement has not just been restricted to visual observations however, with the use of devices which directly measure posture (e.g. goniometers, digitized photographs and radiographic techniques) gaining in popularity. Direct measurement devices have acted to complement, rather than replace, visual observations. In recent times, there has been a proliferation of direct measurement devices, which have evolved to become more sophisticated, accurate and precise. However, this evolution has incurred some costs, as contemporary measurement devices are more expensive and time consuming to use, require a high level of tester expertise, are less versatile, more regionally specific, and in some cases, involve unnecessary health risks (Winkel and Mathiassen, 1994).

Advances in technology have led to recent developments in three-dimensional (3D) analysis, with 3D whole body scanners providing new and exciting opportunities for postural measurement (Buxton *et al.*, 2000). Whole body scanners capture a 3D image of a solid body (e.g. a human body) by passing a light source over its surface, and using a series of cameras, recording the reflected light. The body is then represented as a "point cloud", which is sewn together to create a "digital statue". Complementary measurement-extraction software allows landmarks to be identified (selected), from which numerous digital measurements can be calculated. While whole body scanners fall victim to many of the disadvantages of other direct measurement devices, they do offer several important advantages to postural measurement. First, they require only 10–15 seconds to digitally capture an image of a body; second, they capture an image of the "whole" body, allowing for the calculation of numerous postural measurements; third, captured images provide an historical, graphical record of an individual's posture; fourth, they enable both graphical and numerical feedback to be given; and fifth, there is no exposure to ionizing radiation.

However, even when armed with direct measurement devices, comparisons between postural measurements are largely incommensurable, due to existing methodological differences. So while direct postural measurements are desired, postural comparisons will remain useless, unless an accepted standardized method of postural measurement is adopted. Unfortunately, there is no universally-accepted method of measuring posture (McEvoy and Grimmer, 2005). Therefore, there is a pressing need to describe a method of directly measuring posture – a method which can be used with 3D analysis and other direct measurement devices, and in the clinical setting with visual observation.

This study is the first of a series of studies examining the use of 3D analysis in standing postural measurement. The aim was to describe an anthropometric method of directly measuring standing posture with 3D analysis. The next section describes this anthropometric method by identifying, defining, describing, operationalizing and interpreting a set of standing postural measurements. In addition, the reliability of the method is reported. The final section concludes and makes recommendations to facilitate future comparisons between postural measurements.

2 MEASURING STANDING POSTURE

A systematic strategy was adopted to describe an anthropometric method of directly measuring standing posture with 3D analysis (Figure 5.1). The adopted strategy is described in more detail in the following sections, but briefly, an extensive review of literature was initially conducted to identify studies which explicitly reported on standing postural measurement. Identified studies were reviewed with two questions in mind: "How is posture defined?" and "How are the postural definitions described?" After deciding on one postural description for each of the identified postural definitions, all postural descriptions were operationalized as relationships between landmarks, with the landmarks combining to form the Postural Landmark Set (PLS). Standing postural measurements were then quantified using landmarks from the PLS as inputs into calculations which were created using Cartesian coordinate geometry and trigonometry.

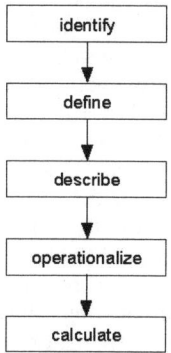

Figure 5.1 Systematic strategy used to describe an anthropometric method of directly measuring standing postural measurements for 3D analysis.

2.1 Identification of postural measurements

The first step in creating an anthropometric method of directly measuring standing posture with 3D analysis, was to try to identify all available measurements used to assess standing posture (Figure 5.1). To do this, an extensive review of the literature was conducted to identify studies which explicitly reported on standing postural measurement. Studies were identified first by a computer search of online bibliographic databases (Cumulative Index to Nursing and Allied Health Literature (CINAHL), the Cochrane Library, and Medline) covering the allied and general health, biomedical, and nursing literature since 1966. The following keywords were used: posture, standing, forward head, round shoulders, kyphosis, lordosis, pelvic tilt, genu valgum, genu varum, patellar alignment, Q-angle, pigeon toes, and in-toes. These keywords were used in combination with the following modifiers:

measurement, assessment and evaluation. Second, the reference lists of all identified studies were manually examined and cross-referenced, with all relevant reports followed up. Studies were excluded if they did not report on static standing postural measurement or were not published in English.

Fifty studies reporting on standing postural measurement were analyzed. The vast majority of these studies (94 per cent or 47 of 50) were peer-reviewed scientific journal articles, with the remainder being published textbooks (4 per cent or 2 of 50) and a post-graduate thesis (2 per cent or 1 of 50). For each of these studies, the following data were recorded: the postural measurement(s), the definition of the postural measurement(s), the description of the postural measurement(s), the landmark(s) used to describe the postural measurement(s), and the measurement device(s) used. The list of studies can be found in Shaw (2005).

Broadly speaking, standing posture is measured either globally as the vertical alignment (in both the Sagittal and Coronal planes) of various body segments (e.g. head, thorax, abdomen, pelvis, thigh and leg), or regionally as individual postural measurements (both within and between body segments). Given the scant literature on the former, and a plethora on the latter, only individual standing postural measurements were considered in this study.

Thirty standing postural measurements were identified from the literature. Of these, 18 were excluded because they (a) were identified only once; (b) were ambiguously defined and described; or (c) could not be measured using 3D analysis. The 12 standing postural measurements retained for this study include all body segments, and are shown in Table 5.1.

To assess the clinical relevance of these 12 standing postural measurements, a questionnaire asking "Which postural measurement(s) do you observe?" was distributed to 40 experienced clinicians (registered clinicians with a minimum of two years full-time, or equivalent part-time, musculoskeletal experience) residing in the Adelaide metropolitan area. Despite a low response rate (38 per cent or 15 of 40), and the majority (87 per cent or 13 of 15) of responses coming from clinicians in only one discipline (physiotherapy), most of the 12 postural measurements (83 per cent or 10 of 12) were reportedly observed by all respondents. The remaining two postural measurements were reportedly observed by all but one respondent. These results confirm at least that these 12 standing postural measurements are clinically relevant.

2.2 Definition of postural measurements

With 12 standing postural measurements identified, the next step was to define them. Most studies described rather than defined standing postural measurements, with few published definitions available. Published definitions were located for 75 per cent (9 of 12) of the standing postural measurements, and were similar among studies. All published definitions were retained for use in this study. Clinical definitions were used for those postural measurements for which a published definition could not be located. Table 5.1 defines each of the 12 standing postural measurements, which are grouped according to the reference plane in which they are measured.

Table 5.1 The definition, description, operationalization and interpretation of each of the 12 identified standing postural measurements.

ID	Postural measurement	Definition	Description	Landmarks	Interpretation
Sagittal plane					
1	Frankfort plane	Degree to which the head is anteriorly or posteriorly tilted	Included angle formed between the intersection of a line joining the lowest point on the inferior margin of the orbit and the tragus of the ear, and a projected line on a Transverse plane (at the level of the tragus of the ear) joining the tragus of the ear and a point on a line projected perpendicularly from the plane to the lowest point on the inferior margin of the orbit	Infraorbitale R, Tragus R	A positive angle indicates that the head is anteriorly tilted, with a smaller angle indicating that the head is less tilted
2	Forward head*	Degree to which the head is anteriorly drawn	Included angle formed between the intersection of a line joining the tragus of the ear and the spinous process of C7, and a projected line on a Transverse plane (at the level of the spinous process of C7) joining the spinous process of C7 and a point on a line projected perpendicularly from the plane to the tragus of the ear	Tragus R, C7	A smaller angle indicates that the head is more anteriorly drawn
3	Round shoulders*	Degree to which the shoulders are anteriorly drawn	Included angle formed between the intersection of a line joining the spinous process of C7 and the acromion, and a projected line on a Transverse plane (at the level of the acromion) joining the acromion and a point on a line projected perpendicularly from the plane to the spinous process of C7	C7, Acromiale R, Acromiale L	A smaller angle indicates that the shoulders are more anteriorly drawn
4	Thoracic kyphosis*	Degree of posterior curvature of the	Included angle formed between the intersection of a line joining the	C7, T3, T7, T12	A smaller angle indicates a greater posterior curvature

#	Parameter	Description	Definition	Landmarks	Interpretation
			spinous process of C7 to the peak of the thoracic spinal curve, and a line joining the peak of the thoracic spinal curve to the spinal process of T12		of the thoracic spine
5	Lumbar lordosis*	Degree of anterior curvature of the lumbar spine	Included angle formed between the intersection of a line joining the spinous process of T12 to the peak of the lumbar spinal curve, and a line joining the peak of the lumbar spinal curve to the spinal process of S2	T12, L3, S2	A smaller angle indicates a greater anterior curvature of the lumbar spine
6	Pelvic tilt	Degree to which the pelvis is anteriorly or posteriorly tilted	Included angle formed between the intersection of a line joining PSIS and ASIS, and a projected line on a Transverse plane (at the level of ASIS) joining ASIS and a point on a line projected perpendicularly from the plane to PSIS	ASIS R, PSIS R	A positive angle indicates that the pelvis is anteriorly tilted, with a smaller angle indicating that the pelvis is less tilted
7	Knee flexion/extension*	Degree to which the knee is flexed or extended	Included angle formed between the intersection of a line joining the greater trochanter of the femur to the lateral epicondyle of the femur, and a line joining the lateral epicondyle of the femur to the lateral malleolus of the fibula, when viewed laterally	Greater trochanter R, Greater trochanter L, Femoral epicondyle lateral R, Femoral epicondyle lateral L, Malleolus lateral R, Malleolus lateral L	An angle >180° indicates that the knee is flexed, with an angle closer to 180° indicating that the knee is less bent

Coronal plane

#	Parameter	Description	Definition	Landmarks	Interpretation
8	Head alignment	Degree to which the head is laterally tilted	Included angle formed between the intersection of a line joining the lowest points on the inferior margins of the left and right orbits, and a projected line on a Transverse plane (at the level of the lowest point on the inferior margin of the right orbit) joining the lowest point on the inferior margin of the right orbit and a point on a line projected perpendicularly from the plane to the lowest point on the	Infraorbitale R, Infraorbitale L	A positive angle indicates that the head is laterally tilted downward to the left, with a smaller angle indicating that the head is less tilted

#	Measure	Description	Landmarks	Interpretation
9	Shoulder alignment	Degree to which the shoulders are bilaterally tilted	Acromiale R, Acromiale L	Included angle formed between the intersection of a line joining the left and right acromia, and a projected line on a Transverse plane (at the level of the right acromion) joining the right acromion and a point on a line projected perpendicularly from the plane to the left acromion	A positive angle indicates that the shoulders are laterally tilted downward to the left, with a smaller angle indicating that the shoulders are less tilted
10	Pelvis alignment	Degree to which the pelvis is laterally tilted	ASIS R, ASIS L	Included angle formed between the intersection of a line joining the left and right ASISs, and a projected line on a Transverse plane (at the level of the right ASIS) joining the right ASIS and a point on a line projected perpendicularly from the plane to the left ASIS	A positive angle indicates that the pelvis is laterally tilted downward to the left, with a smaller angle indicating that the pelvis is less tilted
11	Quadriceps angle*	Degree to which the quadriceps pull superiorly and laterally	ASIS R, ASIS L, Mid-patella R, Mid-patella L, Tibial tuberosity R, Tibial tuberosity L	Included angle formed between the intersection of a line joining ASIS to the midpoint of the patella, and the extension of a line joining the midpoint of patella to the midpoint of the tibial tuberosity	A smaller angle indicates that the pull of the quadriceps is directed more superiorly
12	Genu valgum/varum*	Degree of lateral (valgum) or medial (varum) angulation of the tibia in relation to the femur	Greater trochanter R, Greater trochanter L, Femoral epicondyle lateral R, Femoral epicondyle lateral L, Malleolus lateral R, Malleolus lateral L	Included angle formed between the intersection of a line joining the greater trochanter of the femur to the lateral epicondyle of the femur, and a line joining the lateral epicondyle of the femur to the lateral malleolus of the fibula, when viewed anteriorly	An angle >180° indicates a medial (varum) angulation of the tibia in relation to the femur, with an angle closer to 180° indicating that the tibia is less medially or laterally deviant

R = right; L = left; ASIS = Anterior superior iliac spine; PSIS = Posterior superior iliac spine; C7 = 7th cervical vertebra; T3 = 3rd thoracic vertebra; T7 = 7th thoracic vertebra; T12 = 12th thoracic vertebra; L3 = 3rd lumbar vertebra; S2 = 2nd sacral vertebra. *denotes postural measurements calculated in absolute angular degrees (°)

2.3 Description of postural measurements

Following the generation of a set of postural definitions, a corresponding set of postural descriptions was created. In contrast to the postural definitions, there was no consensus on how standing postural measurements were described. For some postural measurements, only a single description was identified, whereas as for others, there were multiple descriptions. For example, the Forward head (FH) posture is most commonly described as the craniovertebral angle, which is the included angle formed between the intersection of a line joining the tragus of the ear and the spinous process of the seventh cervical vertebra (C7), and a projected line on a Transverse plane (at the level of the spinous process of C7) joining the spinous process of C7 and a point on a line projected perpendicularly from the plane to the tragus of the ear (Braun and Amundson, 1989; Johnson, 1998; Raine and Twomey, 1994; Refshauge *et al.*, 1994). However, FH is also described as the horizontal distance from C7 to the superior tip of the helix of the ear (Grimmer, 1996), and the horizontal distance between the anterior-inferior aspect of the zygomatic bone and the most anterior aspect of the sternal notch (Hickey *et al.*, 2000).

Postural descriptions were chosen using the following criteria. First, if only a single postural description was identified for a postural measurement, then that description was retained; second, if multiple descriptions were identified, then the description which was most commonly reported, or the description thought best for measurement with 3D analysis, was retained. This meant that only one description existed for each postural definition, and hence, each postural measurement. For Thoracic kyphosis and Lumbar lordosis, none of the identified descriptions were appropriate for measurement with 3D analysis, and therefore, a novel description and calculation procedure was created (see Section 2.5.3). Complementing the written postural descriptions in Table 5.1, Figures 5.2 and 5.3 graphically illustrate the postural descriptions in the Sagittal and Coronal planes, respectively.

2.4 Operationalization of postural descriptions

The fourth step to quantify postural measurements was to "operationalize" all postural descriptions as relationships between landmarks (Figure 5.1). The landmarks – identifiable skeletal or soft tissue points on the human body – combined to form the PLS. Table 5.1 identifies the landmarks specific to each postural measurement; Table 5.2 describes the PLS landmarks; and Figures 5.4 and 5.5 show the PLS landmarks relative to the human skeleton. Landmark identification numbers shown in Table 5.2 and Figures 5.4 and 5.5 are in descending order from the superior to the inferior of the human body, along first the anterior and posterior axial skeleton, and second, the anterior and posterior appendicular skeleton.

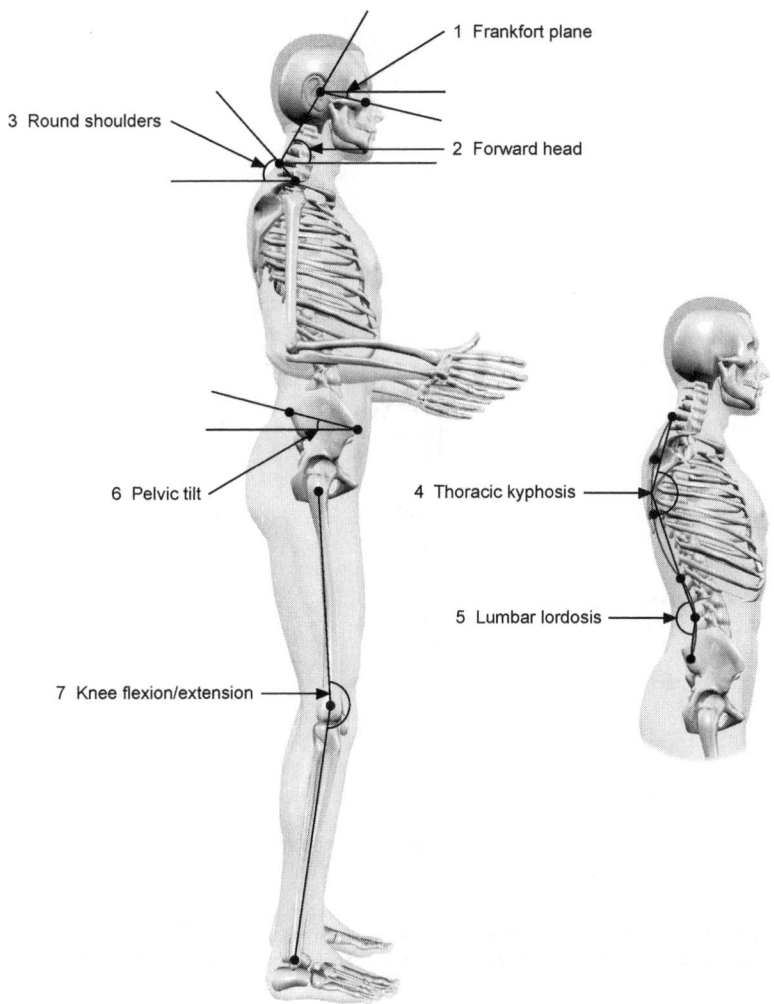

1 Frankfort plane

3 Round shoulders

2 Forward head

6 Pelvic tilt

4 Thoracic kyphosis

5 Lumbar lordosis

7 Knee flexion/extension

Figure 5.2 Lateral view of the body showing the postural measurements measured in the Sagittal plane. The arrows point to the identified measurements. A portion of the skull has been removed to expose the ear, and therefore to assist with viewing the Frankfort plane and Forward head measurements. Thoracic kyphosis and Lumbar lordosis are shown separate to the other postural measurements to better show the spinal landmarks. The spinous processes of the 3rd thoracic and 2nd sacral vertebrae are shown relative to their approximate positions on the spine, despite actually being hidden (in the side view) behind the right scapula and right ilium bones, respectively. The grey spinal curves approximate the quadratic and cubic polynomials used to identify a peak in the thoracic and lumbar spinal curves. The numbers refer to the identification numbers shown in Table 5.1.

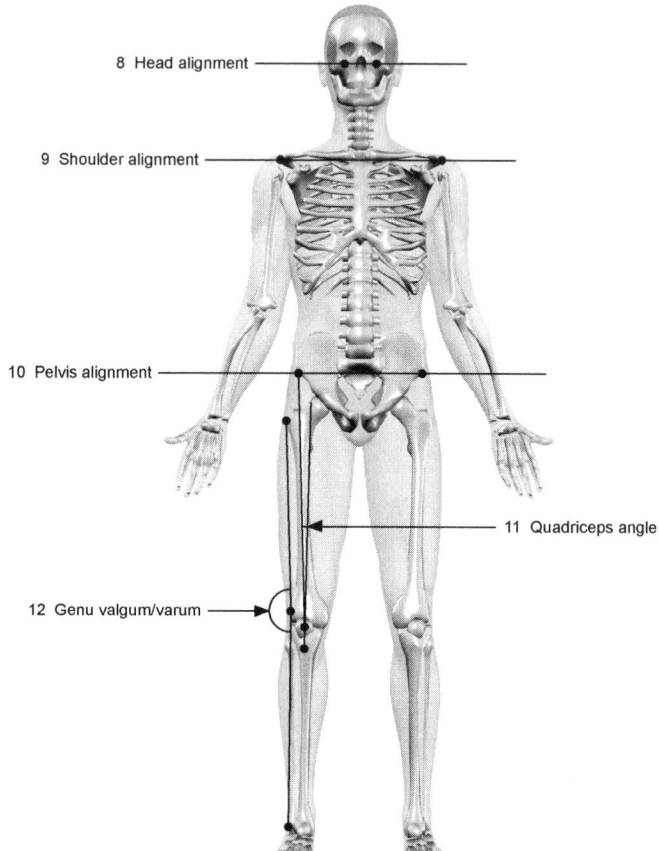

8 Head alignment

9 Shoulder alignment

10 Pelvis alignment

11 Quadriceps angle

12 Genu valgum/varum

Figure 5.3 Anterior view of the body showing the postural measurements measured in the Coronal plane. The arrows point to the identified measurements. The numbers refer to the identification numbers shown in Table 5.1.

The PLS comprises 25 landmarks, 23 of which are skeletal points lying close the body's surface, and two of which (landmark identification numbers 3 and 4 – see Figure 5.4) are soft tissue sites lying on the body's surface. All landmarks are capable of being measured with 3D analysis. Fifty-six percent (14 of 25) of the PLS landmarks were described using clinical descriptions or descriptions identified from the scientific literature. When these descriptions were ill-described or unavailable, descriptions endorsed by the Civilian American and European Surface Anthropometry Resource (CAESAR) (Blackwell *et al.*, 2002) or the International Society for the Advancement of Kinanthropometry (ISAK: International Society for the Advancement of Kinanthropometry, 2001) were used. CAESAR and ISAK descriptions were chosen because they are landmarking systems commonly used throughout the world, and were the systems most familiar to the study authors.

Table 5.2 The description of the landmarks comprising the Postural Landmark Set.

ID	Landmark	Description
Digital		
1,2	Infraorbitale[a]	Lowest point on the inferior margin of the orbit
3,4	Tragus[d]	Mid-point of tragus of the ear
24,25	Malleolus lateral[a]	Most lateral point on the distal fibular protrusion of the ankle
Physical-Digital		
5,6	Acromiale[b]	Most superior and lateral point on the border of the acromion process of the scapula
7,8	ASIS[a]	Most prominent point on the anterior superior spine of the ilium
9	C7	Most prominent posterior point on the spinous process of C7
10	T3	Most prominent posterior point on the spinous process of T3
11	T7	Most prominent posterior point on the spinous process of T7
12	T12	Most prominent posterior point on the spinous process of T12
13	L3	Most prominent posterior point on the spinous process of L3
14	PSIS[a]	Most prominent point on the posterior superior spine of the ilium
15	S2	Most prominent posterior point on the spinous process of S2
16,17	Greater trochanter[d]	Most lateral point on the greater trochanter of the femur
18,19	Femoral epicondyle lateral[a]	Most lateral point on the lateral epicondyle of the femur
20,21	Mid-patella	Mid-point of the anterior surface of the patella
22,23	Tibial tuberosity[c]	Mid-point of tuberosity of the tibia

ASIS = Anterior superior iliac spine; PSIS = Posterior superior iliac spine; C7 = 7th cervical vertebra; T3 = 3rd thoracic vertebra; T7 = 7th thoracic vertebra; T12 = 12th thoracic vertebra; L3 = 3rd lumbar vertebra; S2 = 2nd sacral vertebra.
[a]described by Blackwell *et al.* (2002)
[b]described by International Society for the Advancement of Kinanthropometry (2001)
[c]described by Livinstone and Mandigo (1999)
[d]described by Moore (1992)
Note: bilateral landmarks are identified by two identification numbers (e.g. Infraorbitale). Superscripted letters identify previously described landmarks.

For 3D analysis, PLS landmarks are identified using both physical-digital and digital landmark location; that is, some landmarks need to be marked physically, by attaching an external marker to the human body prior to being identified digitally from the scanned images, while others are obvious enough to be identified digitally (for more details, see Daniell, 2007). The majority of the PLS landmarks were identified using physical-digital location (76 per cent or 19 of 25; Table 5.2). It is recommended that landmarks be identified (and physically marked in the case of physical-digital landmarks) in order of their identification (Table 5.2 and Figures 5.4 and 5.5).

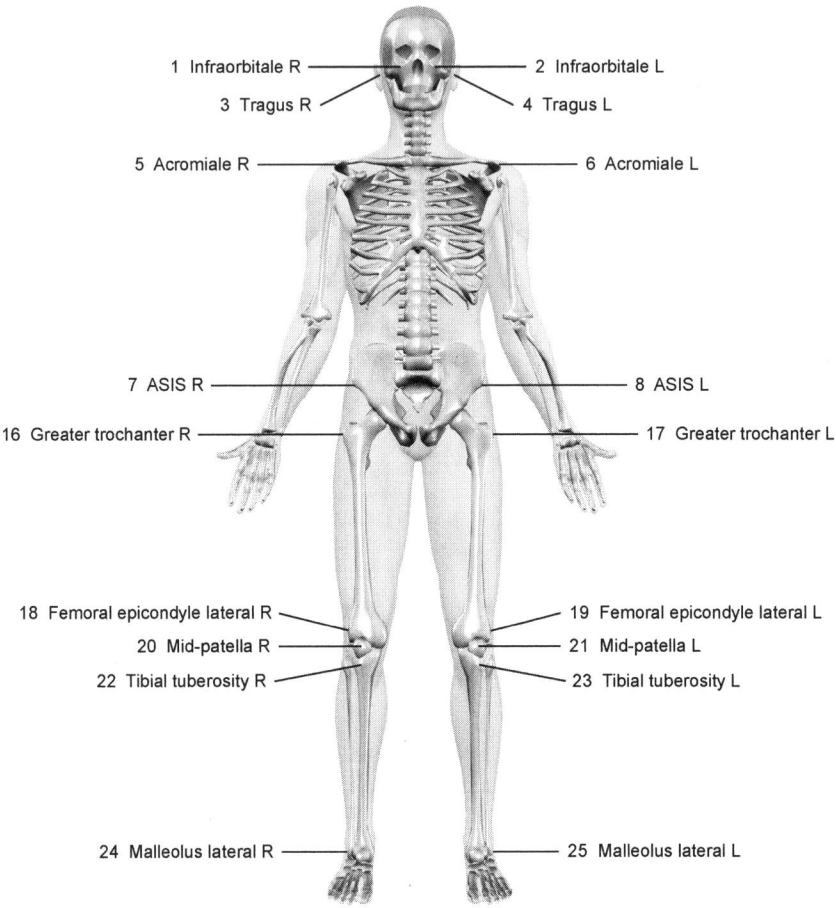

1 Infraorbitale R
2 Infraorbitale L
3 Tragus R
4 Tragus L
5 Acromiale R
6 Acromiale L
7 ASIS R
8 ASIS L
16 Greater trochanter R
17 Greater trochanter L
18 Femoral epicondyle lateral R
19 Femoral epicondyle lateral L
20 Mid-patella R
21 Mid-patella L
22 Tibial tuberosity R
23 Tibial tuberosity L
24 Malleolus lateral R
25 Malleolus lateral L

Figure 5.4 Anterior view of the body showing the landmarks comprising the Postural Landmark Set. "R" is used to denote the right-side of the body, and "L" the left-side.
Note: ASIS = Anterior superior iliac spine.

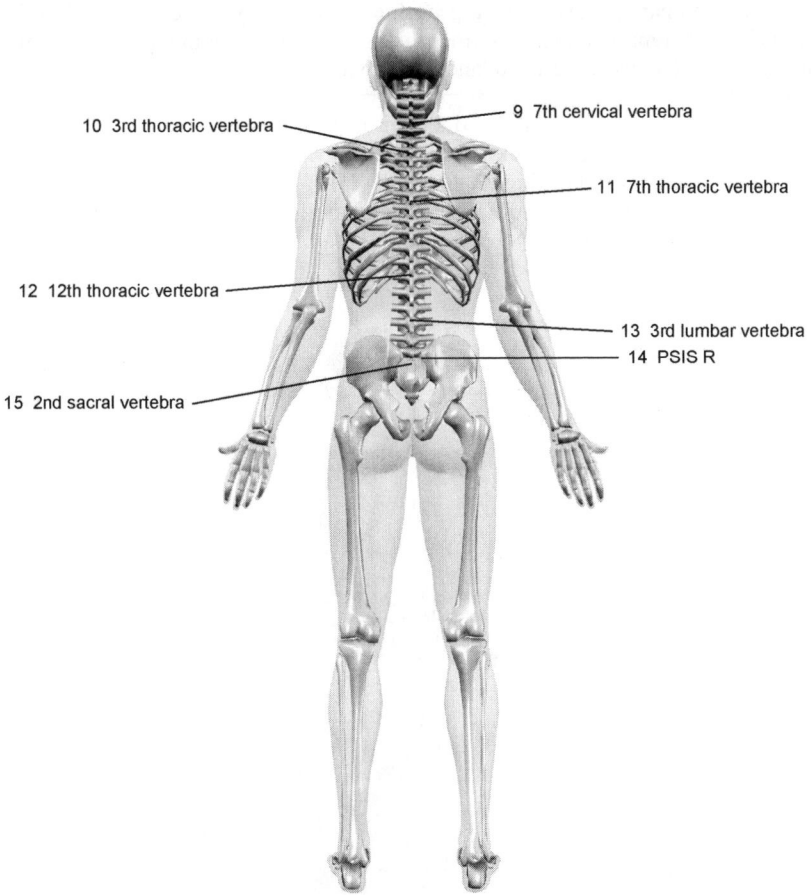

10 3rd thoracic vertebra

9 7th cervical vertebra

11 7th thoracic vertebra

12 12th thoracic vertebra

13 3rd lumbar vertebra

14 PSIS R

15 2nd sacral vertebra

Figure 5.5 Posterior view of the body showing the landmarks comprising the Postural Landmark Set.
"R" is used to denote the right-side of the body, and "L" the left-side.
Note: PSIS = Posterior superior iliac spine.

It is important that subjects be landmarked and scanned in bare feet and in the same "normal" standing posture. Figure 5.6 shows the "normal" standing posture assumed by subjects prior to being physically landmarked and scanned. With the goal to measure static standing posture, the subject's arms should be bent to 90° to allow access (by both the anthropometrist and 3D scanner) to the upper thigh and pelvis. This will assist in reducing movement artefacts likely to affect postural measurements.

The landmarking and scanning posture assumed by subjects should be both "normal" and reproducible. To find their "normal" standing posture, subjects should be verbally asked prior to landmarking and scanning to:

- Take a few steps in place to find their normal foot position;
- Bend their head and neck to look down, then up, then straight ahead to find their normal head position; and
- Bend their elbows to 90° (without moving their shoulders) to allows access to the upper thigh and pelvis.

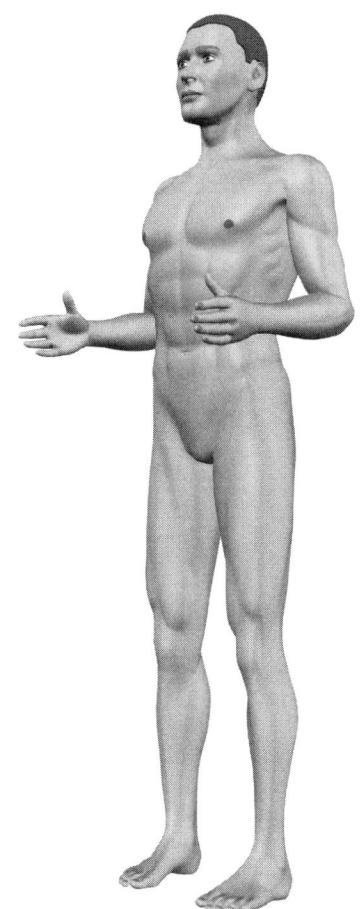

Figure 5.6 The "normal" standing posture assumed by subjects prior to physical landmarking and scanning. The elbows are bent to 90°.

2.5 Calculation of postural measurements

The final step in creating an anthropometric method of directly measuring standing posture with 3D analysis was to calculate the magnitude of each postural measurement (Figure 5.1). A review of the literature revealed that direct measurements of standing posture are principally measured as angles (e.g. the included angle formed between the intersection of a line joining two landmarks and a projected line on a reference plane at the level of one of the landmarks) or linear distances (e.g. the distance between a landmark and a reference plane). Because this study merges 12 standing postural measurements, a consistent method of analysis was used to calculate the magnitude of the postural measurements. All postural measurements were calculated in angular degrees (°) using calculations created using Cartesian coordinate geometry and trigonometry.

Postural measurements were calculated as (a) the included angle formed between the intersection of a line joining two measured landmarks, and a projected line on a reference plane at the level of one of the landmarks (58 per cent or 7 of 12); (b) the included angle formed between the intersection of a line joining two measured landmarks, and a second line joining one of these landmarks to another measured landmark (25 per cent or 3 of 12); and (c) the included angle formed between the intersection of a line joining one measured and one derived landmark, and a second line joining the derived landmark to another measured landmark (17 per cent of 2 of 12) (Table 5.1 and Figures 5.2 and 5.3). Seventy-five percent (8 of 12) of the standing postural measurements consider direction, while the remainder do not.

Table 5.1 describes how each of the 12 standing postural measurements are interpreted. It is important to note that some postural measurements (e.g. Round shoulders, Knee flexion/extension, Quadriceps angle and Genu valgum/varum) can be calculated bilaterally (i.e. on both sides of the body), using landmarks identified on both the right- and left-sides of the body separately. It is therefore recommended that bilateral measurements be taken on both the right- and left-sides of the body (because there may be unilateral deviations, e.g. in the case of stroke), and unilateral measurements be taken only on the right-side of the body.

Through 3D analysis, PLS landmarks are identified by unique Cartesian (XYZ) coordinates, which can be used to quantify the magnitude of standing postural measurements. The steps involved to move from "real" human bodies to XYZ coordinates through 3D analysis have been previously described (Buxton *et al.*, 2000; Daniell, 2007). To better describe the calculation of the standing postural measurements, XYZ coordinates were respectively referred to as FSU coordinates, where F is the front-back coordinate, S the side-side coordinate, and U the up-down coordinate. Standing postural measurements in the Sagittal plane (Table 5.1) require only the F and U coordinates of the corresponding PLS landmarks, and those in the Coronal plane only the S and U coordinates.

2.5.1 Calculation of the included angle formed between the intersection of a line joining two measured landmarks, and a projected line on a reference plane at the level of one of the landmarks

The calculation of an included angle formed between the intersection of a line joining two measured landmarks, and a projected line on a reference plane at the level of one of the landmarks, requires the coordinates of two measured landmarks and one derived landmark. For example, the FH measurement uses the F and U coordinates of the measured Tragus (right) and C7 landmarks, and the F and U coordinates of a derived landmark on a Transverse plane at the level of the C7 landmark. By using the coordinates of three relevant landmarks to construct a right-angled triangle, trigonometry can be used to calculate the corresponding postural measurement.

To illustrate, consider the FH measurement (Table 5.1 and Figure 5.2). FH is described and operationalized as the included angle formed between the intersection of a line joining the Tragus (right) and C7 landmarks, and a projected line on a Transverse plane (at the level of the C7 landmark) joining the C7 landmark and a point on a line projected perpendicularly from the plane to the Tragus (right) landmark. For these postural measurements, the derived landmark is taken to be the F coordinate of the first described landmark [e.g. Tragus (right)] and the U coordinate of the second described landmark (e.g. C7) (see Table 5.1 for the order of the landmark descriptions). Taken together, the two measured landmarks and one derived landmark form a right-angled triangle, with the opposite side formed between the first described landmark [e.g. Tragus (right)] and the derived landmark; the adjacent side between the second described landmark (e.g. C7) and the derived landmark; and the hypotenuse between the first and second described landmarks [e.g. Tragus (right) and C7, respectively] (Figure 5.7). With only the lengths of the opposite and adjacent sides required, the length of the opposite side is calculated as the U coordinate of the first described landmark [e.g. Tragus (right)] minus the U coordinate of the derived landmark, and the length of the adjacent side calculated as the F coordinate of the second described landmark (e.g. C7) minus the F coordinate of the derived landmark (Figure 5.7). These side lengths are then substituted into Equation 1 to calculate the angle describing the postural measurement.

$$\theta = \tan^{-1}\left(\frac{\text{opposite}}{\text{adjacent}}\right) \cdot \left(\frac{180}{\pi}\right) \qquad (5.1)$$

Where, θ = the angle (°) describing the postural measurement; opposite = the length (mm) of the opposite side; and adjacent = the length (mm) of the adjacent side.

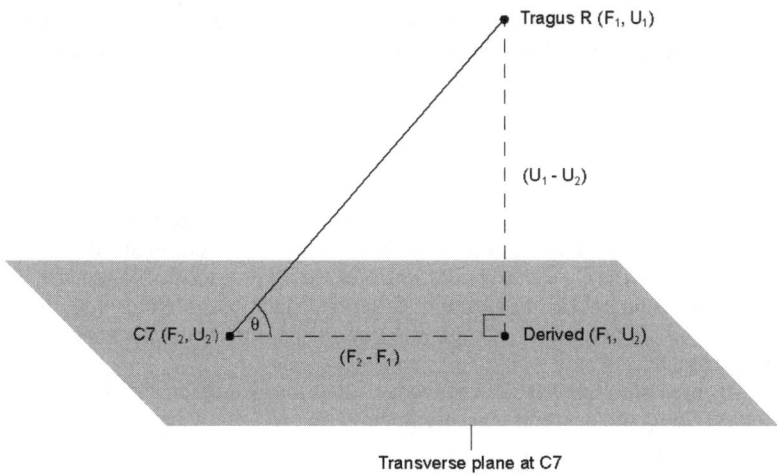

Figure 5.7 An illustration of the coordinate data required to calculate the Forward head measurement. The included angle is identified as θ, the opposite side length as (U_1-U_2), and the adjacent side length as (F_1-F_2). *Note*: Tragus R = Tragus (right) and C7 = 7th cervical vertebra.

2.5.2 Calculation of the included angle formed between the intersection of a line joining two measured landmarks, and a second line joining one of these landmarks to another measured landmark

The calculation of an included angle formed between the intersection of a line joining two measured landmarks, and a second line joining one of these landmarks to another measured landmark, requires the coordinates of three measured landmarks. For example, the Knee flexion/extension (right) measurement uses the F and U coordinates of the measured Greater trochanter (right), Femoral epicondyle lateral (right) and Malleolus lateral (right) landmarks. By using the coordinates of three relevant measured landmarks to construct a triangle, Cartesian coordinate geometry and trigonometry can be used to calculate the corresponding postural measurement.

To illustrate, consider the Knee flexion/extension (right) measurement (Table 5.1 and Figure 5.2). When the leg is viewed laterally, Knee flexion/extension is described and operationalized as the included angle formed between the intersection of a line joining the Greater trochanter (right) and Femoral epicondyle lateral (right) landmarks, and a line joining the Femoral epicondyle lateral (right) and the Malleolus lateral (right) landmarks. Taken together, these three measured landmarks form a triangle, with side *a* formed between the first and second described landmarks [e.g. Greater trochanter (right) and Femoral epicondyle lateral (right), respectively]; side *b* between the second and third described landmarks [e.g. Femoral epicondyle lateral (right) and Malleolus lateral (right), respectively]; and side *c* between the first and third described landmarks [e.g. Greater trochanter (right) and Malleolus lateral (right), respectively] (see Table 5.1 for the order of the landmark descriptions). The side lengths (*a*, *b* and *c*) are calculated using Equation 5.2.

$$\text{length} = \sqrt{\left(F_1 - F_2\right)^2 + \left(U_1 - U_2\right)^2} \tag{5.2}$$

Where, length = the length (mm) between two landmarks; F_1 = the F coordinate of the first described landmark; F_2 = the F coordinate of the second described landmark; U_1 = the U coordinate of the first described landmark; and U_2 = the U coordinate of the second described landmark. Note, for the Genu valgum/varum and Quadriceps angle measurements, the F coordinates are replaced by S coordinates.

All three side lengths are then substituted into Equation 5.3 to calculate the angle describing the postural measurement.

$$\theta = \cos^{-1}\left(\frac{a^2 + b^2 - c^2}{2ab}\right) \cdot \left(\frac{180}{\pi}\right) \tag{5.3}$$

Where, θ = the angle (°) describing the postural measurement; a = the length (mm) of the side between the first and second described landmarks; b = the length (mm) of the side between the second and third described landmarks; and c = the length (mm) of the side between the first and third described landmarks. Note, for the Quadriceps angle measurement, the resultant angle needs to be subtracted from 180°.

2.5.3 Calculation of the included angle formed between the intersection of a line joining one measured and one derived landmark, and a second line joining the derived landmark to another measured landmark

Because published descriptions for Thoracic kyphosis and Lumbar lordosis were inappropriate for measurement with 3D analysis, a novel description (Table 5.1) and calculation procedure was created. The calculation of an included angle formed between the intersection of a line joining one measured and one derived landmark, and a second line joining the derived landmark to another measured landmark, requires the coordinates of two measured landmarks and those of one derived landmark. For example, the Lumbar lordosis measurement uses the F and U coordinates of the measured T12 and S2 landmarks, and the F and U coordinates of a derived landmark at the peak of the lumbar (lower) spinal curve. By using the coordinates of three relevant landmarks to construct a triangle, Cartesian coordinate geometry and trigonometry can be used to calculate the corresponding postural measurement.

To illustrate, consider the Lumbar lordosis measurement (Table 5.1 and Figure 5.2). Lumbar lordosis is described and operationalized as the included angle formed between the intersection of a line joining the T12 landmark and the peak of the lumbar spinal curve, and a line joining the peak of the lumbar spinal curve to the S2 landmark. The derived landmark is taken to be F and U coordinates of the peak of the spinal curve. For example, the measured F and U coordinates of the T12, L3 and S2 landmarks mark the curvature of the lumbar spine (note, the measured F and U coordinates of the C7, T3, T7 and T12 landmarks mark the curvature of the thoracic spine). The coordinates of the peak of the spinal curve are determined using the D_{max} procedure (see Cheng *et al.*, 1992). D_{max} works by fitting a curve (usually a quadratic or cubic polynomial) to a set of coordinates (in this case spinal coordinates). Note, for this procedure to work, the spinal coordinates need to be swapped (i.e. the F becomes the U, and the U becomes the F) prior to curve fitting. A quadratic polynomial is used when three F and U coordinates are available (e.g. Lumbar lordosis), with a cubic polynomial used when four or more F and U coordinates are available (e.g. Thoracic kyphosis). A straight line is then drawn between the coordinates of the end-landmarks of the spinal curve (e.g. the F and U coordinates of T12 and S2). D_{max} occurs at the greatest perpendicular distance between the fitted polynomial curve and its corresponding straight line, and is calculated mathematically by rotating the polynomial curve until the straight line is horizontal, and finding the minimum point on the curve by differentiation. This procedure derives an F coordinate at D_{max}, and by substituting back into the polynomial regression equation, a corresponding U coordinate, and hence, a spinal curve peak.

Taken together, the measured end-landmarks, and the D_{max}-derived landmark, form a triangle, with side *a* formed between the first described end-landmark (e.g. T12) and the D_{max}-derived landmark; side *b* between the D_{max}-derived landmark and the second end-described landmark (e.g. S2); and side *c* between the first and second described end-landmarks (e.g. T12 and S2, respectively) (see Table 5.1 for the order of the landmark descriptions). The side lengths are calculated using Equation 5.2, and then substituted into Equation 5.3, to calculate the angle describing the postural measurement.

2.6 Reliability of postural measurements using 3D analysis

The reliability of this anthropometric method for 3D postural analysis has recently been determined by Shaw (2005). Using a convenience sample of 52 asymptomatic South Australian adults aged 18–62 years, intra-rater reliability was examined by comparing duplicate measurements by the same researcher taken 24 hours apart. Bland and Altman (1986) analysis was used to quantify intra-rater reliability. Systematic bias was calculated as the mean difference between duplicate measurements, with positive biases indicating smaller second measurements and negative biases smaller first measurements. Random error was calculated as the 95 per cent "limits of agreement" (i.e. 1.96 times the standard deviation of the differences).

Table 5.3 shows the results of the intra-rater reliability analysis by Shaw (2005). Only 13 per cent (2 of 15) of postural measurements – Frankfort plane and Forward head – were 95 per cent likely to be systematically biased. The median bias was –0.1° (range –3.4° to 2.9°). Random error was greatest for postural measurements involving the head and neck (i.e. Forward head and Frankfort plane).

Table 5.3 The intra-rater reliability analysis by Shaw (2005). Shown are the means (SD) for the first and second postural measurements, the mean (95 per cent CI) and standard deviation (95% CI) of the measurement differences, and the absolute "limits of agreement". All measurements are shown in degrees (°).

Postural measurement	Scan 1 Mean (SD)	Scan 2 Mean (SD)	Difference Mean (95% CI)	Difference SD (95% CI)	Absolute limits
Forward head	53.6 (8.7)	57.0 (6.7)	−3.4 (−6.9 to −0.0)	12.6 (9.2 to 16.0)	−3.4 (24.6)
Frankfort plane	8.2 (8.1)	5.3 (4.9)	2.9 (0.2 to 5.5)	9.8 (7.2 to 12.5)	2.9 (19.3)
Quadriceps angle (right)	13.8 (6.5)	14.0 (6.9)	−0.2 (−1.7 to 1.2)	5.3 (3.9 to 6.8)	−0.2 (10.4)
Lumbar lordosis	158.3 (8.8)	159.2 (9.3)	−0.9 (−2.3 to 0.5)	5.1 (3.7 to 6.5)	−0.9 (10.0)
Quadriceps angle (left)	11.1 (6.5)	11.6 (6.8)	−0.6 (−1.9 to 0.8)	5.0 (3.7 to 6.4)	−0.6 (9.9)
Round shoulders (right)	50.5 (12.6)	50.3 (12.3)	0.2 (−1.1 to 1.5)	4.6 (3.4 to 5.9)	0.2 (9.1)
Thoracic kyphosis	152.4 (5.6)	153.1 (6.2)	−0.7 (−1.8 to 0.4)	3.9 (2.8 to 5.0)	−0.7 (7.6)
Pelvic tilt	10.6 (5.7)	10.0 (5.9)	0.6 (−0.3 to 1.6)	3.4 (2.5 to 4.4)	0.6 (6.7)
Head alignment	−0.8 (2.3)	−1.2 (2.5)	0.4 (−0.4 to 1.2)	2.9 (2.1 to 3.8)	0.4 (5.8)
Knee flexion/extension (right)	176.1 (2.8)	176.2 (2.7)	−0.1 (−0.8 to 0.6)	2.6 (1.9 to 3.3)	−0.1 (5.0)
Knee flexion/extension (left)	175.9 (3.1)	176.2 (2.9)	−0.3 (−0.8 to 0.3)	2.1 (1.5 to 2.7)	−0.3 (4.1)
Pelvis alignment	−0.2 (2.2)	−0.4 (1.8)	0.2 (−0.3 to 0.7)	1.7 (1.2 to 2.2)	0.2 (3.3)
Shoulder alignment	−0.5 (1.6)	−0.8 (1.7)	0.3 (−0.0 to 0.6)	1.1 (0.8 to 1.5)	0.3 (2.2)
Genu valgum/varum (right)	176.4 (2.6)	176.3 (2.7)	0.1 (−0.2 to 0.4)	1.1 (0.8 to 1.5)	0.1 (2.2)
Genu valgum/varum (left)	175.6 (2.9)	175.5 (2.8)	0.1 (−0.3 to 0.4)	1.1 (0.8 to 1.5)	0.1 (2.2)

95% CI = 95 per cent confidence interval.
SD = standard deviation.
Note: reliability data are not available for Round shoulders (left).

Using the within-subjects standard deviation *vs.* the between-subjects standard deviation as a criterion for "acceptable" reliability (i.e. "acceptable" if the former is less than the latter), Shaw (2005) concluded that all postural measurements, except Forward head and Frankfort plane, were acceptably reliable.

3 CONCLUSION AND RECOMMENDATIONS

Over the years, standing posture has been measured in many different ways. Without a universally-accepted method of measuring standing posture, comparisons between postural measurements are difficult, if not impossible, due to methodological differences. This study is unique, as it describes an anthropometric method of directly measuring standing posture with 3D analysis – a method which is capable of being adapted for use with other direct measurement devices and visual observation. The described anthropometric method has many advantages for postural measurement. First, it was built on a systematic review of all published reports explicitly commenting on standing postural measurement; second, it defines, describes, operationalizes and interprets a set of regionally-based postural measurements, which are inclusive all of the body's segments, and can combine to give a "global" assessment of standing posture; and third, it comprises clinically relevant postural measurements. To facilitate future comparisons between postural measurements, the following recommendations are made:

- Methodological differences have led to postural measurements being largely incommensurable. A single measurement protocol, such as that described in this study, should be used, or the protocol used should be accurately reported.
- Regardless of the measurement protocol used, researchers should define, describe, operationalize and interpret all postural measurements. The measurement device(s) used should also be reported.
- When having standing posture measured, it is important that individuals are in bare feet and assume a "normal" posture, for both landmarking and scanning (or physical measuring). Researchers are encouraged to use clear instructions to assist individuals assume their "normal" posture.
- All postural landmarks should be clearly identified prior to postural measurement(s), and researchers are encouraged to rigorously practise both physical and digital landmarking.
- Bilateral postural measurements should be taken on both the right- and left-sides of the body, and unilateral postural measurements taken only on the right-side of the body.
- Postural measurements should be reported in angular degrees (°).
- Factors which are likely to affect the day-to-day variability of standing posture (e.g. time of measurement and physical activity) should be controlled and/or reported.
- Finally, all postural measurements should be accompanied by an estimate of precision, so that single measurements can be reported with confidence and changes over time tracked.

ACKNOWLEDGEMENTS

We would like to thank Maureen McEvoy, Ian Fulton, and Tim Olds for their helpful advice on quantifying standing posture.

REFERENCES

Arnold, C., Beatty, B., Harrison, E. and Olszynski, W., 2000, The reliability of five clinical postural alignment measures for women with osteoporosis. *Physiotherapy Canada*, **2**, 286–294.

Blackwell, S., Robinette, K.M., Bœhmer, M., Fleming, S., Kelly, S., Brill, T., Hœferlin, D., Burnsides, D. and Daanen, H., 2002, *Civilian American and European Surface Anthropometry Resource (CAESAR), Final Report, Volume II: Descriptions* (Warrendale, PA: States Air Force Research Laboratory).

Bland, M. and Altman, D., 1986, Statistical methods for assessing agreement between two methods of clinical measurement. *Lancet*, **1**, 307–310.

Braun, B.L. and Amundson, L.R., 1989, Quantitative assessment of head and shoulder posture. *Archives of Physical Medicine and Rehabilitation*, **70**, 322–329.

Buxton, B., Dekker, L., Douros, I. and Vassilev, T., 2000, Reconstruction and interpretation of 3D whole body surface images. In *Proceedings of the International Conference of Numerisation 3D-Scanning* (Paris).

Cheng, B., Kuipers, H., Snyder, A.C., Keizer, H.A., Jeukendrup, A. and Hesselink, M., 1992, A new approach to the determination of ventilatory and lactate thresholds. *International Journal of Sports Medicine*, **13**, 518–522.

Daniell, N., 2007, A comparison of the accuracy of the Vitus Smart and Hamamatsu Body Line 3D whole-body scanners. In: Marfell-Jones, M. and Olds, T. (Eds). *Kinanthropometry X. Proceedings of the 9th Conference of the International Society for the Advancement of Kinanthropometry (ISAK).* (London: Routledge).

Fedorak, C., Ashworth, N., Marshall, J. and Paull, H., 2003, Reliability of the visual assessment of cervical and lumbar lordosis: How good are we? *Spine*, **28**, 1857–1859.

Grimmer, K., 1996, The relationship between cervical resting posture and neck Pain. *Physiotherapy*, **82**, 45–51.

Hickey, E.R., Rondeau, M.J., Corrente, J.R., Abysalh, J. and Seymour, C., 2000, Reliability of the Cervical Range of Motion (CROM) device and plumb-line techniques in measuring Resting Head Posture (RHP). *Journal of Manual and Manipulative Therapy*, **8**, 10–17.

International Society for the Advancement of Kinanthropometry, 2001, *International Standards for Anthropometric Assessment* (Potchestroom: International Society for the Advancement of Kinanthropometry).

Johnson, G.M., 1998, The correlation between surface measurement of head and neck posture and the anatomic position of the upper cervical vertebrae. *Spine*, **23**, 921–927.

Livinstone, L.A. and Mandigo, J.L., 1999, Bilateral Q angle asymmetry and anterior knee pain syndrome. *Clinical Biomechanics*, **14**, 7–13.

McEvoy, M. and Grimmer, K., 2005, Reliability of upright posture measurements in primary school children, *BMC Musculoskeletal Disorders*, vol. 6. Online. Available HTTP: <http://www.biomedcentral.com/1471-2474/6/35> (accessed 25 July 2005).

Moore, K.L., 1992, *Clinically Oriented Anatomy*, 3rd ed. (Baltimore: Williams and Wilkins).

Raine, S. and Twomey, L., 1994, Attributes and qualities of human posture and their relationship to dysfunction or musculoskeletal pain. *Critical Reviews in Physical and Rehabilitation Medicine*, **6**, 409–437.

Refshauge, K., Goodsell, M. and Lee, M., 1994, Consistency of cervical and cervicothoracic posture in standing. *Australian Journal of Physiotherapy*, **40**, 235–240.

Shaw, L.G., 2005, *The Quantification of Static Standing Posture Using 3D Imaging*, unpublished thesis, University of South Australia.

Tyson, S., 2003, A systematic review of methods to measure posture. *Physical Therapy Reviews*, **8**, 45–50.

Winkel, J. and Mathiassen, S.E., 1994, Assessment of physical work load in epidemiologic studies: Concepts, issues and operational considerations. *Ergonomics*, **37**, 979–988.

CHAPTER SIX

Revising sizing: Modifying clothing templates to match the 3D shape of real women

J. Fraser and T. Olds

University of South Australia

1 INTRODUCTION

Women commonly report difficulty in finding clothing that fits them well, largely because the size and shape assumed by recommended clothing templates do not match the body shapes of real women today. The aim of this study was to develop size-shape templates for garments which provided a better fit than existing templates.

This study compared the 3D shapes assumed by Standards Australia (SA) size-shape templates to the 3D shapes of 294 18–30 year-old, tertiary-educated Australian women, measured using 3D whole-body scanning. Each woman was assigned a best-fit size from the SA sizing system using a dimension-matching algorithm, with each of the 28 SA dimensions weighted according to subjective discomfort. The average measurement for each dimension was calculated across all women assigned to each best-fit size category. These measurements were then used as the new dimensional standards for that size category. Women were again assigned to their (new) best-fit size. This process was repeated until there were no further statistically significant reductions in across-sample lack of fit.

2 DISSATISFACTION WITH FIT

Dissatisfaction with the fit of off-the-rack clothing has been widely reported, especially amongst women, with common findings indicating that at least 50% of the female population experience difficulty finding well-fitting clothing for the size and shape of their bodies (Anderson *et al.*, 2001; Anderson-Connell *et al.*, 2002; Otieno *et al.*, 2005; Simmons and Istook, 2003; Sproles and Geistfeld, 1978). This leads to the need for costly alterations, in addition to a loss of profit for retailers and manufacturers due to unsold and returned items, which ultimately increases the cost of goods to the consumer (Otieno *et al.*, 2005).

3 STANDARDS AUSTRALIA SIZING SYSTEM

3.1 Origins

The industry standards for defining clothing sizes in Australia were first developed by Standards Australia (SA) based on anthropometric data collected in 1926. Whilst the standards were updated in 1969, the revised standards were based on self-report data of a limited number of measurements (Berlei, 2005; Standards Australia, 1997).

3.2 Mismatches with real women

There is evidence that the actual size and shape of Australian women today does not match with the size and shape of bodies assumed by the SA sizing system. In particular, the SA system has unrealistically low waist-hip ratios. Using data from the Australian Anthropometric Database (AADBase) and from NHANES III data on US adults, we can demonstrate that the actual size and shape of Australian women today does not match with that assumed by the Standards Australia (SA) sizing system. When individuals from the samples are matched to their best fit size based on their hip girth, WHR is larger in the population sample for all sizes.

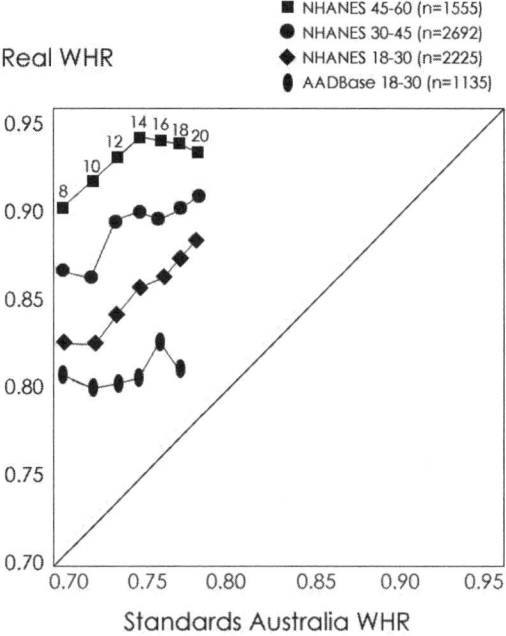

Figure 6.1 Waist to hip ratios of women from the AADBase and NHANES III compared to waist to hip ratios assumed by Standards Australia templates.

4 SECULAR TRENDS

This difference between the outdated sizing system and data on real women is due, in part, to the fact that the size and shape of the population have been changing for at least several generations.

4.1 Height and relative leg length

Current and continuing secular trends in increasing height have been extensively reported around the world, with research indicating current growth rates of greater than 1 cm per decade (Arcaleni, 2006; Bielicki and Szlarka, 1999; Castilho and Lahr, 2001; Damon, 1968; Eiben *et al.*, 2005; Fredriks *et al.*, 2000; Freedman *et al.*, 2000; Kautiainen *et al.*, 2002; Kuh *et al.*, 1991; Lahti-Koski *et al.*, 2000; Liese *et al.*, 2001; Liestol and Rosenberg, 1995; Lissner *et al.*, 1998; Monteiro *et al.*, 1994; Olivier, 1980; Padez and Johnston, 1999; Rasmussen *et al.*, 1999; Rosengren *et al.*, 2000; Silventoinen *et al.*, 1999; Simmons *et al.*, 1996; Smith and Norris, 2004; Torrace *et al.*, 2002; Ulijaszek, 2001).

The secular increase in height is occurring mainly due to an increase in leg length, indicating that relative body proportions are also changing (Ayub Ali *et al.*, 2000; Bogin *et al.*, 2002; Jantz and Jantz, 1999; Roche, 1979; Sanna and Soro, 2000; Tanner *et al.*, 1982).

4.2 Overweight and obesity

Extensive research into increasing prevalence of overweight and obesity has found secular increases in weight, body mass index (BMI) and waist-to-hip ratio (WHR), also suggesting change in body morphology (Castilho and Lahr, 2001; Damon, 1968; Eiben *et al.*, 2005; Kautiainen *et al.*, 2002; Kuskowska-Wolk and Bergstrom, 1993a,b; Lahti-Koski *et al.*, 2000; Liese *et al.*, 2001; Liestol and Rosenberg, 1995; Lissner *et al.*, 1998; Rasmussen *et al.*, 1999; Rosengren *et al.*, 2000; Seidell *et al.*, 1995; Shimokata *et al.*, 1989; Simmons *et al.*, 1996; Torrace *et al.*, 2002; Ulijaszek, 2001; Yoshike *et al.*, 2002).

4.3 Body Mass Index

In Figure 6.2, data from 25 population studies on the secular increase in BMI are presented. Only two population groups showed a decrease in BMI over time. Each data point on the graph represents the average BMI of the sample at that point in time as a percentage of the average BMI for that sample in 1992. When data was not collected for a study in 1992, I interpolated or extrapolated the estimated average BMI in that year based on the rate of increase in BMI per year. A steady increase in BMI is evident in the population since the 1960s. Given the assumption that weight is proportional to the square of height, this increase in BMI implies that the human form must have undergone transformations in its relative proportions in order to accommodate for this increase in body mass, relative to height.

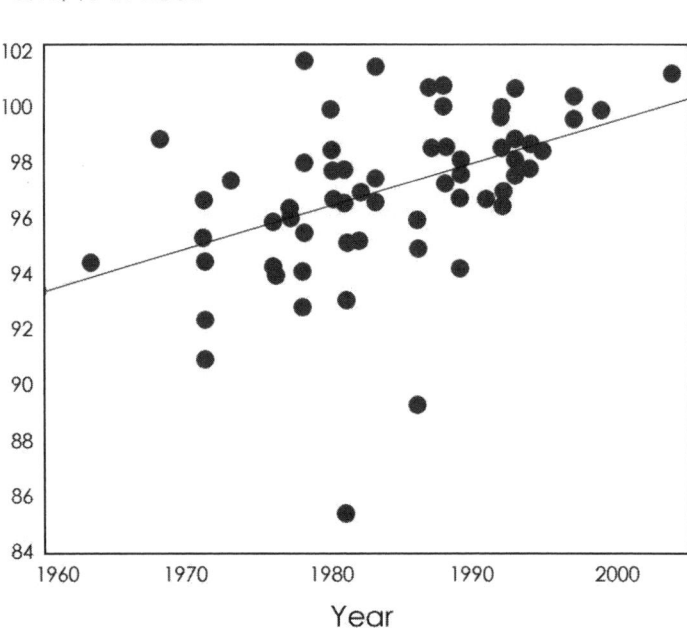

BMI, % of 2000

Year

Figure 6.2 Increase in average BMI in 25 population samples over time. There has been an increase in average BMI of approximately 7% in the past four decades.

4.4 Waist-to-hip ratio

In line with a secular increase in BMI, is a trend towards increased relative waist girth compared to hip girth, due to increased abdominal adiposity. This change in relative proportions results in women today having a generally more "apple-shaped" figure, compared to the "pear-shaped" figure defined in the Standards Australia sizing system.

Again, ample data are available to indicate a secular increase in WHR. Additionally, the available data indicate that the increase in WHR is due almost solely to an increase in waist circumference, and that hip circumference has showed virtually no change over time. Figure 6.3 shows the secular increases in WHR in the time period between 1968 to 2000.

WHR, % of 1992

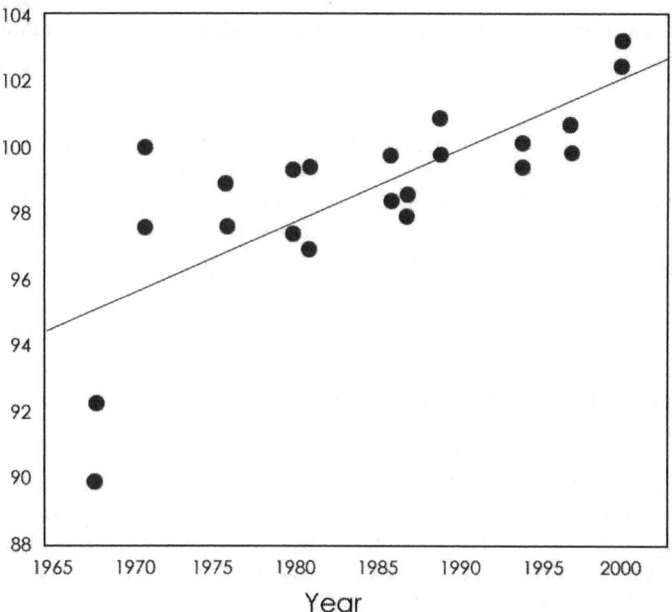

Figure 6.3 Increase in average WHR in 11 population samples over time. There has been an increase in average WHR of approximately 8% in the past four decades.

5 VANITY SIZING

Perhaps because of the fact that the current sizing system doesn't fit women today, many manufacturers choose to develop their own size shape templates for garment production. Additionally, manufacturers and retailers use what is known as "vanity sizing," where a garment is labelled as a smaller size than if the sizing system were adhered to. This practice aims to appeal to the egos of consumers and help create customer loyalty. The result is that the value of numerical sizing has become virtually obsolete (Anderson-Connell *et al.*, 2002).

6 THREE DIMENSIONAL ANTHROPOMETRY

There is a clear need for an up-to-date sizing system to be developed. In order for this to occur, a sufficiently large anthropometric survey needs to occur to obtain statistically useful data on the size and shape of Australian women today. Three dimensional whole body scanning technology is a means to allow this to occur.

6.1 Advantages of 3D scanning

Three-dimensional whole body scanning captures a representation of the topography and geometry of the body in a fast and reproducible way (Daanen and Van De Water, 1998). From the 3D scan it is possible to obtain a potentially infinite number of linear and non-linear measurements (Istook and Hwang, 2001). Importantly, once collected, data can be stored and then retrieved and referred to at any time following the initial data collection, and new information can be extracted (Daanen and Van De Water, 1998). This makes it practical for data to be collected from a large sample group to enable development of a sizing system that more accurately fits the population group (Treleaven, 2004).

7 ADAPT PROJECT

The data used were originally collected for use in the Australian Defence Anthropometric Personnel Testing (ADAPT) project, conducted in 2005 by the University of South Australia, in conjunction with the Australian Defence Force (ADF). The ADAPT project collected anthropometric data from Royal Australian Airforce aircrew and age-matched civilians using 3D whole body scanning technology. 3D whole body scanners captured three-dimensional images of participants. Measurement extraction software was used to derive anthropometric data from these images. These data are to be used to direct the design of future aircraft crewstations and revise recruitment standards.

7.1 Measurement error

Three-dimensional anthropometry has been found to have acceptably low measurement error when compared to traditional anthropometry. In the ADAPT project, participants were physically landmarked prior to scanning and landmarks manually located on the computer by a trained operator. Systematic error was assessed using bias, and random error was assessed using the standard error of measurement (SEM). SEM is the standard deviation of the differences divided by $\sqrt{2}$. When compared to physical measurement extraction, measurements derived from the scanning system had an overall SEM of 5.1 mm, with a bias of 5.2 mm, which was considered to be within acceptable limits (Rogers, 2005).

8 METHOD

8.1 Subject selection

Subjects were recruited on a voluntary basis from randomly-selected regions in Australia. Recruitment criteria required subjects aged between 18 and 30 years old, holding Australian citizenship and currently completing, or having completed tertiary study. The data used in this study were a random subset of the female data from the ADAPT data. A sample of 294 women, aged between 18 and 30 years were included. An additional set of data was collected from a separate holdback sample with the same characteristics as the main sample. This sample was used to test whether a reduction in lack of fit for the main sample between the original and the developed sizing system would correspond to a similar reduction in lack of fit in the population.

8.2 Scanning surveys

Scans of participants were obtained during scanning surveys at various Australian university campuses and Air Force bases during 2005, as part of the ADAPT project. All scanning survey procedures are outlined in the ADAPT Procedures Manual (Olds *et al.*, 2005). Prior to scanning, subjects were landmarked by a qualified anthropometrist who placed physical landmarkers on 22 defined anatomical locations, or landmarks. Subjects were scanned in the standard scanning posture for the ADAPT project, a neutral standing posture, using the Vitus Smart laser scanner.

8.3 Measurement extraction

Following scan processing and file conversions, anthropometric measurements were extracted from BAF files, using the measurement extraction software CySize. This was performed as part of the previous study by Honey (2007), which compared the body shapes assumed by the SA sizing system with a sample of the population of young Australian women. The measurements extracted correspond to the 28 dimensions in the SA sizing system, consisting of lengths, widths, breadths, and girths.

This process required an operator to manually select the landmarkers on each scan. These corresponded to the anatomical locations used to define measurements in the SA sizing system. The operator then chose a measurement from the pre-programmed set of measurements, and selected the appropriate landmarks corresponding to that measurement. The software calculated a distance for that measurement, following the convex, but not the concave, contours of the body, as is the case when distances are measured with a tape measure in traditional anthropometry.

8.4 Lack of fit (L) statistic

The previous study quantified the lack of fit between body shapes assumed by the SA sizing system and the actual body shapes and sizes of young Australian women. This involved determining the size within the SA sizing system that is each subject's "best fit." This was done using the lack of fit (L) statistic, which was devised specifically for this project:

Given

1   A set of dimensions (1 ... i ... n) representing the n dimensions which make up each size-shape template. These are the 28 dimensions defining each size-shape template in the SA sizing system.

2   A series of size shape templates (1 ... s ... p) representing the p size-shape templates in the sizing system. These are the nine sizes of the SA sizing system (8 to 24).

3   A series of subjects (1 ... j ... m) representing the m subjects in the sample.

Let

4   S_{ils} be the value of the ith dimension of size–shape template s.

5   M_{ij} be the measurement for the ith dimension on the jth subject.

Then

6        $|M_{ij} - S_{ils}|$ represents the absolute difference (in cm) between the value of the ith dimension of size-shape template s, and the corresponding measurement on subject j.

7        $|M_{ij} - S_{ils}| W_i$ represents the subjectively weighted difference. A survey was conducted on an independent sample of 50 women to determine a weighting for each dimension by getting subjects to rate how uncomfortable or unacceptable they find an item of clothing to be if a particular dimension of the garment is too big or too small. There is a separate weighting for whether the garment is too big or too small.

8        $L_{jls} = [\sum_{i=1}^{i=n}(|M_{ij} - S_{ils}| W_i)]/i$ represents the overall lack of fit for subject j relative to size-shape template s – i.e. the average of all the lack of fit values for each of the n dimensions.

9        $minL_j = \min_{s=1}^{s=p} (L_{jls})$ is the lowest L-value for subject j relative to any of the p size-shape templates in the sizing system. This is the best-fit size.

10       $L_{tot} = (\sum_{j=1}^{j=m} minL_j)/m$ is the overall lack of fit for the entire sample of m subjects relative to the sizing system.

8.5 Developing new size-shape templates

Lack of fit (L) values were calculated for each subject for each size in the original Standards Australia sizing system, which we denoted as SA1. Participants were allocated to the size with the least L value, known as their "best fit size." This process resulted in sample groups for each best fit size.

From here, for each "best fit" size sample, we took the average of the subjects' measurements for each dimension. These formed the values for each dimension in the new size-shape template for each size. This new sizing system was called SA2.

The new values for the dimension in each size meant that some subjects would now have a new best fit size. The L-statistic was again applied to each subject to determine the size in the new sizing system which was their new best-fit size. This gave us new sample groups for each best fit size. Again, we determined the overall lack of fit for these new size-shape templates with the new best-fit size samples.

When this resulted in a significant improvement in lack of fit, we repeated the outlined process in an attempt to minimise lack of fit. Again, averages were calculated for each dimension in the new size samples, subjects reallocated to "best-fit" sizes, and overall lack of fit calculated for the new sizing system. The subsequent sizing system was known as SA3. We continued this process until no further significant reduction in lack of fit was achieved. This resulted in our final sizing system, SA4.

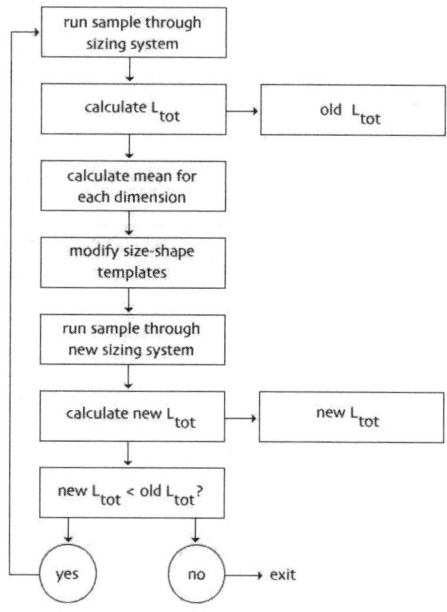

Figure 6.4 Flowchart of data processing in this study.

9 RESULTS

This process was applied to data on 294 female subjects aged 18–30, obtained from the ADAPT project. After the initial allocation of women to their best fit size in the Standards Australia sizing system, the total lack of fit for the data with respect to the sizing system was found to be 19.81. The mean values for each dimension for women in each best fit size were then used to form new size shape templates. When women were again allocated to their best fit size, the new lack of fit for the sizing system was found to be 13.42, and on a second and third run through it further reduced to 13.20 and 13.02. This represented a significant reduction of 33% of the total lack of fit values for the sizing system.

Table 6.1 Reduction in lack of fit (L_{tot}) for the n = 294 sample.

Sizing System	Mean	SD	Median	Min	Max
SA1	19.81	6.00	18.60	9.9	61.0
SA2	13.42	4.94	12.40	5.0	51.1
SA3	13.20	4.70	12.35	5.3	45.8
SA4	13.02	4.49	12.20	5.4	43.7

In order to verify that the sizing system developed was capable of significantly reducing lack of fit in the population and not just in the sample that the new sizing system was based on, we calculated lack of fit of a separate holdback sample of 75 subjects from the ADAPT project. For this sample, the overall L value for the original standards Australia sizing system was slightly lower at 16.6. The lack of fit of this sample with the SA4 sizing system was 13.02, which represented a significant reduction in lack of fit.

Table 6.2 Reduction in lack of fit (L_{tot}) for the n = 75 holdback sample.

Sizing System	Mean	SD	Median	Min	Max
SA1	16.66	4.83	15.10	8.9	33.8
SA2	13.12	3.90	12.40	8.0	26.4
SA3	13.19	3.93	12.40	8.1	28.4
SA4	13.20	3.93	12.30	8.0	28.5

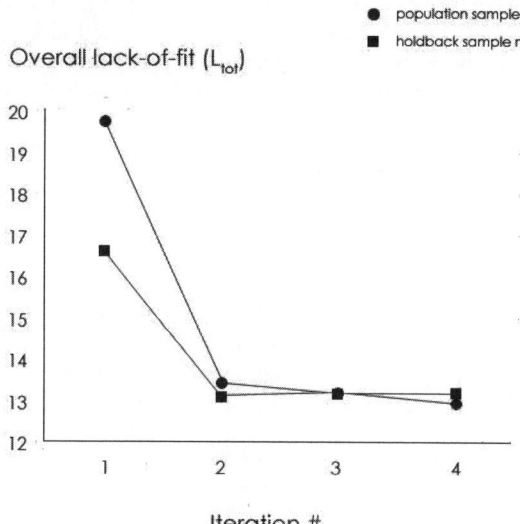

Figure 6.5 Decrease in across sample lack of fit between the original Standards Australia sizing system and the new sizing systems developed.

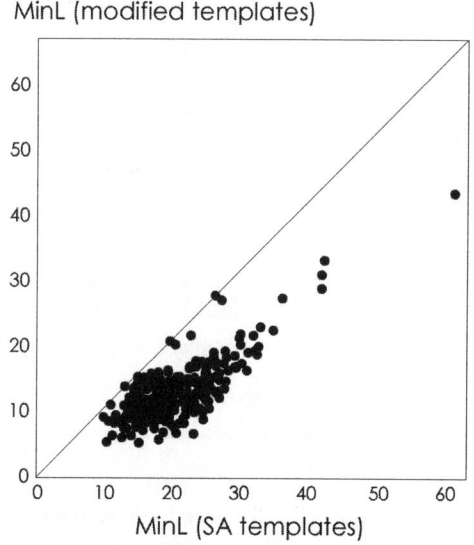

Figure 6.6 Change in minimum lack of fit for individual subjects between the original Standards Australia sizing system and the SA4 sizing system.

For all measurements except two, there was a decrease in lack of fit between SA1 to SA4. The greatest improvement in lack of fit occurred in knee height, hip height, underbust girth and waist height, which all improved by greater than 50%. Hip girth and arm length lack of fit values increased. This variation in the change in L values indicates that it may be necessary for some dimensions to have a greater degree of lack of fit in order to achieve an overall decrease in lack of fit.

Table 6.3 Change in lack of fit values of dimensions between SA1 and SA4.

Dimension	SA1	SA4	SA4% of SA1
knee height	38.05	8.95	23.5
hip height	54.07	21.19	39.2
underbust girth	40.76	18.79	46.1
waist height	42.34	19.91	47.0
inside leg length	40.02	20.46	51.1
neck-base girth	18.60	9.65	51.9
elbow girth	6.65	3.51	52.8
knee girth	11.02	7.00	63.5
knee height	15.54	10.47	67.4
cervical height	28.18	19.56	69.4
armscye to waist	12.18	8.86	71.3
shoulder length	6.61	4.81	72.8
upper arm girth	11.37	8.35	73.4
abdominal extension	27.24	20.21	74.2
bust girth	19.35	14.43	74.6
across chest width	13.71	10.62	77.5
armscye girth	16.74	13.23	79.0
back waist length	11.24	9.32	82.9
neck shoulder point to breast point	9.28	7.74	83.4
back width	11.32	9.71	85.8
underarm length	15.12	13.08	86.5
waist to hips	16.92	15.05	88.9
waist girth	15.66	14.14	90.3
total crotch length	20.17	18.43	91.4
bust width	7.60	7.19	94.6
thigh girth	13.58	12.88	94.8
hip girth	17.2	20.35	118.3
arm length	14.27	17.05	119.5

The change in actual values for some individual dimensions will now be considered. One of the dimensions which had a large decrease in lack of fit was hip height. As depicted in Figure 6.7, left panel, for sizes 8 to 20 the spread of values was much greater in SA4 than in SA1. For sizes 8 and 10, SA1 hip height is larger than SA4, while for sizes 12 to 20, SA1 hip height is smaller than SA4. Whilst hip height for sizes 8 to 20 increase systematically, sizes 22 and 24 do not fit this pattern. This reflects the fact that larger individuals are not necessarily taller. The values for hip height were more spread out in SA4 than SA1. The original Standards Australia system did not adequately account for variation in hip height amongst the population.

Between SA1 and SA4, lack of fit values for waist girth decreased by 10%. Figure 6.7, right panel, shows that there does not appear to be an overall pattern occurring to explain how lack of fit has been improved. SA4 values for waist girth did not vary greatly from SA1, indicating that original values for waist girth were fairly accurate, compared to the original values for hip height, which were too closely clustered together in the original templates.

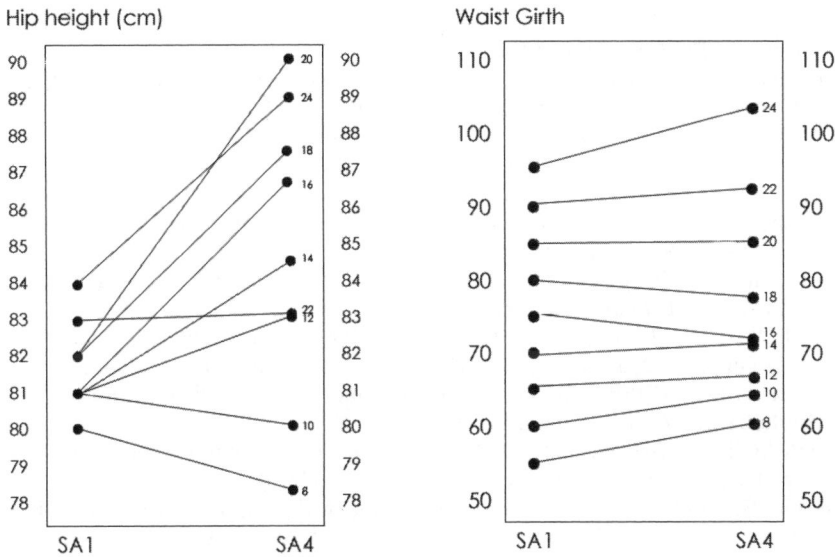

Figure 6.7 Change in dimension values for hip height and waist girth between SA1 and SA4 for sizes 8 to 24.

10 DISCUSSION

A sizing system has been developed that has a significantly reduced lack of fit in 18–30 year old women, compared to the SA sizing system. Furthermore, a process has been established for developing better-fitting sizing systems given any original sizing system and anthropometric data on a corresponding population sample.

The next step in developing a sizing system that more accurately fits the size and shape of the population is to start with new templates. These could be based on a more elaborate, multi-dimensional model, including the use of angles, volumes, and cross-sectional areas. A new sizing system could better reflect the great variation of shapes and sizes in the population by having a greater number of variables, compared to the current "step-ladder" system, which assumes that all dimensions get bigger or smaller simultaneously. This could occur by grouping individuals in a sample into shape clusters and developing sizes that reflect not only the size, but also the common shapes in the population.

11 ACKNOWLEDGEMENTS

This project was funded in part by the ADF Project M15872.

12 REFERENCES

AADBase, 1995, "Australian Anthropometric Database," School of Health Sciences, The University of South Australia.

Anderson, L.J., Brannon, E.L., Ulrich, P.V., Presley, A.B., Woronka, D., Grasso, M. and Stevenson, D., 2001, *Understanding Fitting Preferences of Female Consumers: Development of an Expert System to Enhance Accurate Sizing Selection*. National Textile Centre Annual Report.

Anderson-Connell, L.J., Ulrich, P.V. and Brannon, 2002, "A consumer-driven model for mass customization in the apparel market," *Journal of Fashion Marketing & Management*, **6**(3), 240–258.

Arcaleni, E., 2006, "Secular trend and regional differences in the stature of Italians, 1854–1980," *Economics and Human Biology*, **4**(1), 24–38.

Ayub Ali, M.D., Uetake, T. and Ohtsuki, F., 2000, "Secular changes in relative leg length in post-war Japan," *American Journal of Human Biology*, **12**, 405–416.

Berlei, 2005, History of the Berlei Group, http://www.berlei.com.au/About-Berlei/History/Time-Line.asp.

Bielicki, T. and Szklarska, 1999, "Secular trends in stature in Poland: national and social class-specific," *Annals of Human Biology*, **26**(3), 251–258.

Bogin, B., Smith, P., Orden, A.B., Varela Silva, M.I. and Loucky, J., 2002, "Rapid change in height and body proportions of Maya American children," *American Journal of Human Biology*, **14**, 753–761.

Castilho, L.V. and Lahr, M.M., 2001, "Secular trends in growth among urban Brazilian children of European descent," *Annals of Human Biology*, **28**(5), 564–574.

Daanen, H.A.M. and Van De Water, G.J. 1998, "Whole body scanners," *Displays*, **19**(3), 111–120.

Damon, A., 1968, "Secular trend in height and weight within old American families at Harvard, 1870–1965 within four generation families," *American Journal of Physical Anthropology*, **29**, 45–50.

Eiben, G., Dey, D.K., Rothenberg, E., Steen, B., Bjorkelund, C., Bengtsson, C. and Lissner, L., 2005, "Obesity in 70-year-old Swedes: secular changes over 30 years," *International Journal of Obesity*, **29**, 810–817.

Fredriks, A.M., Van Buuren, S., Burgmeijer, R.J.F., Meulmeester, J.F., Beuker, R.J., Brugman, E., Roede, M.J., Verloove-Vanhorik, S.P. and Wit, J.-M., 2000,

"Continuing positive secular growth change in the Netherlands 1955–1997," *Pediatric Research*, **47**, 316–323.

Freedman, D.S., Kettel Khan, L., Serdula, M.K., Srinivasan, S.R., and Berenson, G.S., 2000, "Secular trend in height among children during 2 decades," *Archives of Paediatrics & Adolescent Medicine*, **154**(2), 155–161.

Honey, F. and Olds, T., 2007, The Standards Australia Sizing System: Quantifying the Mismatch. In: Marfell-Jones, M.J. and Olds, T. (Eds) *Kinanthropometry X. Proceedings of the 10th Conference of the International Society for the Advancement of Kinanthropometry (ISAK)*. (London: Routledge).

Istook, C.L. and Hwang, S.J., 2001, "3D body scanning systems with application to the apparel industry", *Journal of Fashion Marketing and Management*, **5**(2), 120–132.

Jantz, L.M. and Jantz, R.L., 1999, "Secular change in long bone length and proportion in the United States, 1800–1970," *American Journal of Physical Anthropology*, **110**(1), 57–67.

Kautiainen, S., Rimpela, A., Vikat, A., and Virtanen, S.M., 2002, "Secular trends in overweight and obesity among Finnish adolescents in 1977–1999," *International Journal of Obesity*, **26**, 544–552.

Kuh, D.L., Power, C. and Rodgers, B., 1991, "Secular Trends in Social Class and Sex Differences in Adult Height," *International Journal of Epidemiology,* **20**(4), 1001–1009.

Kuskowska-Wolk, A. and Bergström, R., 1993a, "Trends in body mass index and prevalence of obesity in Swedish men 1980–1989," *Journal of Epidemiology and Community Health*, **47**, 103–108.

Kuskowska-Wolk, A. and Bergström, R., 1993b, "Trends in body mass index and prevalence of obesity in Swedish women 1980–1989," *Journal of Epidemiology and Community Health*, **47**, 195–199.

Lahti-Koski, M., Pietinen, P., Mannisto, S. and Vartiainen, E., 2000, "Trends in waist-to-hip ratio and its determinants in adults in Finland from 1987 to 1997," *The American Journal of Clinical Nutrition*, **72**(6), 1436–1444.

Liese, A.D., Doring, A., Hense, H.W. and Keil, U., 2001, "Five year changes in waist circumference, body mass index and obesity in Augsburg, Germany," *European Journal of Nutrition*, **40**(6), 282–288.

Liestol, K. and Rosenberg, M., 1995, "Height, weight and menarcheal age of schoolgirls in Oslo – an update," *Annals of Human Biology*, **22**(3), 199–205.

Lissner, L., Bjorkelund, C., Heitmann, B.L., Lapidus, L., Bjorntorp, P. and Bengtsson, C., 1998, "Secular increases in waist-hip ratio among Swedish women," *International Journal of Obesity,* **22**(11), 1116–1120.

Loesch, D.Z., Stokes, K. and Huggins, R.M., 2000, "Secular trend in body height and weight of Australian children and adolescents," *American Journal of Physical Anthropology*, **111**, 545–556.

Monteiro, C.A., D'Aquino Benicio, M.H. and Da Cruz Gouveia, N., 1994, "Secular growth in Brazil over three decades," *Annals of Human Biology*, **21**(4), 381–390.

Olds, T., Tomkinson, G., Rogers, M., Kupke, T., Lowe, L., Daniell, N. and Honey, F., 2005, *ADAPT Procedures Manual*, The University of South Australia, Adelaide.

Olivier, G., 1980, "The increase of stature in France," *Journal of Human Evolution*, **9**, 645–649.

Otieno, R., Harrow, C. and Lea-Greenwood, G., 2005, "The unhappy shopper, a retail experience: exploring fashion, fit and affordability," *International Journal of Retail & Distribution Management*, **33**(4), 298–309.

Padez, C. and Johnston, F., 1999, "Secular trends in male adult height 1904–1996 in relation to place of residence and parent's educational level in Portugal," *Annals of Human Biology*, **26**(3), 287–298.

Rasmussen, F., Johansson, M. and Hansen, H.O., 1999, "Trends in overweight and obesity among 18-year-old males in Sweden between 1971 and 1995," *Acta Paediatrica*, **88**, 431–437.

Roche, A.F. (ed), 1979, "Secular Trends in Human Growth, Maturation and Development," *Monographs of the Society for Research in Child Development*, 44(3-4), 1–53.

Rogers, M., 2005, "Quantifying error in 3D anthropometry using a Vitus Smart whole body scanner," BAppSc (Hons) thesis, University of South Australia.

Rosengren, A., Eriksson, H., Larsson, B., Svardsudd, K., Tibblin, G., Welin, L. and Wilhelmsen, L., 2000, "Secular changes in cardiovascular risk factors over 30 years in Swedish men aged 50: the study of men born in 1913, 1923, 1933 and 1943," *Journal of Internal Medicine*, **247**(1), 111–118.

Sanna, E. and Soro M.R., 2000, "Anthropometric changes in urban Sardinian children 7–10 years between 1975–1976 and 1996," *American Journal of Human Biology*, **12**, 782–791.

Seidell, J.C., Verschuren, W.M.M. and Kromhout, D., 1995, "Prevalence and trends of obesity in The Netherlands 1987–1991," *International Journal of Obesity*, **19**, 924–927.

Shimokata, H., Andres, R., Coon, P.J., Elahi, D., Muller, D.C. and Tobin, J.D., 1989, "Studies in the distribution of body fat. II. Longitudinal effects of change in weight," *International Journal of Obesity*, **13**, 455–464.

Silventoinen, K., Lahelma, E. and Rahkonen, O., 1999, "Social background, adult body-height and health," *International Journal of Epidemiology*, **28**(5), 911–918.

Simmons, G., Jackson, F., Swinburn, B. and Yee, R.L., 1996, "The increasing prevalence of obesity in New Zealand: Is it related to recent trends in smoking and physical activity?" *New Zealand Medical Journal*, **109**, 90–92.

Simmons, K.P. and Istook, C.L., 2003, "Body measurement techniques, comparing 3D body-scanning and anthropometric methods for apparel applications," *Journal of Fashion Marketing and Management*, **7**(3), 306–332.

Smith, S.A. and Norris, B.J., 2004, "Changes in the body size of UK and US children over the past three decades," *Ergonomics*, **47**(11), 1195–1207.

Sproles, G.B. and Geistfeld, L.V., 1978, "Issues in analysing consumer satisfaction/dissatisfaction with clothing and textiles," *Advances in Consumer Research*, **5**(1), 383–391.

Standards Australia, 1997, *Australian Standard: Size Coding System for Women's Clothing – Underwear, Outerwear and Foundation Garments*, Standards Australia, NSW.

Tanner, J.M., Hayashi, T., Preece, M.A. and Cameron, N., 1982, "Increase in length of leg relative to trunk in Japanese children and adults from 1957 to 1977: A comparison with British and with Japanese Americans," *Annals of Human Biology*, **9**(5), 411–423.

Torrace, G.M., Hooper, M.D. and Reeder, B.A., 2002, "Trends in overweight and obesity among adults in Canada (1970–1992): Evidence from national surveys using measured height and weight," *International Journal of Obesity*, **26**(5), 797–804.

Treleaven, P., 2004, "Sizing us up," *IEEE Spectrum*, **4**(41), 29–31.

Ulijaszek, S.J., 2001, "Increasing body size among adult Cook Islanders between 1966 and 1996," *Annals of Human Biology*, **28**(4), 363–373.

Yoshike, N., Seino, F., Tajima, S., Arai, Y., Kawano, M., Furuhata, T. and Inoue, S., 2002, "Twenty-year changes in the prevalence of overweight in Japanese adults: the National Nutrition Survey 1976–95," *Obesity Reviews*, **3**, 183–190.

CHAPTER SEVEN

The Standards Australia sizing system: Quantifying the mismatch

F. Honey and T. Olds

University of South Australia, Adelaide, Australia

1 INTRODUCTION

Garments produced for the Australian clothing industry are, in principle, manufactured according to the Standards Australia (SA) garment sizing system (Standards Australia, 1997). Despite a sizing system being in place, high levels of dissatisfaction are reported in relation to off-the-rack clothing (Anderson-Connell, Ulrich and Brannon, 2002; Ross, 2003; Simmons and Istook, 2003; Sproles and Geistfeld, 1978; Treleaven, 2004). This chapter aims to explain some of the possible explanations for clothing dissatisfaction and to outline use of the L-statistic, a new method for comparing two sets of measurement data.

2 LITERATURE REVIEW

The SA sizing system consists of a detailed set of measurements for each size category (8-24), such as arm length, waist girth and hip height. When put together, the set of measurements partially specifies the three-dimensional (3D) shape of the body of a woman who would fit perfectly into a garment of each particular size (Standards Australia, 1997). The current sizing system was first developed in 1959, based on the US model and data from Berlei's sizing survey of 1926, the last major anthropometric survey conducted in Australia (Pontoni, 2004). Since this time, it has been reviewed into its current form. However, the last major review was in 1970 and was based on self-reported waist, hip, bust and height measurements (Standards Australia, 1997).

High levels of dissatisfaction with off-the-rack clothing have been reported and may be explained by inaccurate development of the current sizing system (Anderson-Connell, Ulrich and Brannon, 2002; Ross, 2003; Simmons and Istook, 2003; Sproles and Geistfeld, 1978; Treleaven, 2004). However, a more apparent reason is the growing trend of "vanity sizing", where designers purposely label a garment as a smaller size. This is common in more-expensive, higher-end apparel and is seen as a way of gaining a competitive edge (Biederman, 2003).

When comparing waist-hip ratios (WHR) from the Australian Anthropometric database (AADBase) with those found in the Standards Australia size templates, there is a poor concordance, with women from the database generally having lower waist-hip ratios than those found in the size templates. The AADBase is a non-random database of measurements collected by trained anthropometrists. Women in this database have not reported their usual dress size, so they have been matched to clothing sizes according to hip girth (AADBase, 1995). Figure 7.1 below shows the WHR assumed in the SA sizing system (open dots) with those taken from the AADBase (filled dots), based on assigned clothing size. Shape does change with size, but not in the same way as the size templates (Standards Australia, 1997).

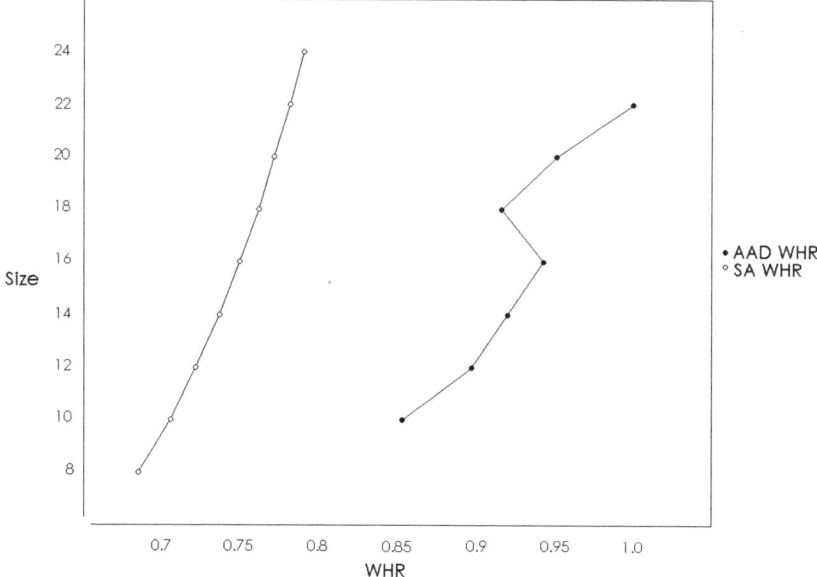

Figure 7.1 Comparison of WHR as assumed by SA templates (open dots), to those found in the Australian Anthropometric Database (filled dots).

Secular trend in height and weight is another reason that there is a mismatch between clothing sizes and Australian women. Since at least the mid 19th century, anthropometrists have noted a steady increase in the height and weight of adults (Cole, 2003). In the past 150 years, the average height of Europeans has increased by more than 20 cm. This trend is evident all over the world (Usher, 1996). Increases in height between generations occur mainly in the first two years of life and are due to an increase in leg length (Cole, 2003). The large changes in leg length mean that our relative proportions are changing. Weight is also reported to be undergoing a secular increase, with an up to 3.4 kg increase in Cook Islanders over a 10-year period (Ulijaszek, 2001).

The increases in height and weight in the past 150 years have not been at the same rate. Weight is increasing much faster, as adult BMI values have also increased over this period (Lahti-Koski *et al.*, 2000). Waist-to-hip ratio (WHR) and waist circumference measures have also increased over time. These variations mean that both body shape and body size are changing (Thomas, 2004).

The secular changes in body size and shape suggest that the current size templates are no longer an accurate representation of our current population (Standards Australia, 1997). Also, the current sizing system does not account for changing body shapes resulting from modern undergarments (such as push up bras, control top underwear, etc.) and ethnic diversity which has increased in Australia due to immigration (Lang, 1995).

This study aimed to analyse the Standards Australia sizing system, to suggest ways in which the current system can be improved. To accomplish this, the L-statistic was developed and applied to a sample of young Australian women. This was done in conjunction with using 3D whole body scanners to describe the shape of real Australian women.

3 METHODOLOGY

To quantify the mismatch between the shapes assumed in the SA sizing system and the real body shapes of Australian women, the "lack-of-fit" or L-statistic was developed. The L-statistic is a way to quantify

- The lack-of-fit for one individual in one dimension, compared to one size template (for example, how well a size 10 fits a particular woman around the hips);
- The overall lack-of-fit for one individual relative to one size template (for example, how well a size 10 fits a particular woman generally);
- The best match between any size template in a sizing system and the size and shape of any individual (for example, which size fits a particular woman best); and
- The overall lack-of-fit for a number of subjects relative to a sizing system (for example, how well the SA sizing system fits 18–30-year-old Australian women).

When devising the L-statistic, it was hypothesised that lack-of-fit for some specific dimensions may be more problematic than in others. For example, a sweater that is too loose around the elbow may cause little discomfort, compared to a pair of pants that is too tight around the waist. Therefore, a questionnaire was developed to determine how subjectively important fit is for each dimension, providing a weighting factor for that specific dimension. This questionnaire was given to an independent sample of 45 18–30-year-old Australian women. Each dimension found in the SA size templates was thereby given a weighting factor, if the dimension was too big, or too small. Table 7.1 outlines these weighting factors. Values in the "too small" column refer to the rating given when the SA dimension is smaller than the physical measure (for example, a sleeve that is too short, or pants that are too tight). The "too big" column refers to ratings given when the SA dimension is larger than the physical measure (for example, pants that are too long, or a t-shirt that is too loose) (Standards Australia, 1997).

Table 7.1 Subjective weighting values for each SA dimension (Standards Australia, 1997). Higher values indicate greater subjective discomfort.

SA dimension	Too small	Too big
Bust girth (bras)	7.7	7.9
Inside leg length	7.0	7.8
Waist girth (pants)	7.5	7.3
Underbust girth	7.2	7.4
Hip girth	7.7	6.5
Waist to hips	6.8	7.4
Total crotch length	7.0	7.2
Bust width	5.8	7.3
Underarm length	8.3	4.2
Abdominal extension	6.5	5.8
Thigh girth	7.3	4.8
Waist height	7.7	4.3
Back waist length	7.6	4.4
Upper arm girth	7.5	4.5
Hip height	8.2	3.7
Armscye girth	7.5	4.3
Waist girth (tops)	7.0	4.7
Arm length	6.9	4.7
Back width	6.9	4.7
Neck-base girth	7.5	4.1
Across chest width	6.5	5.0
Knee height	5.3	6.2
Front waist length	6.3	5.0
Bust girth (tops)	6.7	4.4
Shoulder length	5.9	5.2
Armscye to waist	6.4	4.6
Neck shoulder point to breast point	6.1	4.9
Cervical height	5.7	4.5
Knee girth	4.8	3.3
Elbow girth	3.8	3.5
Overall	6.8	5.3

Following the generation of weighting values for each dimension, the L-statistic was calculated in the following way:

Given

1   A set of dimensions (1 ... i ... n) representing the n dimensions which make up each size-shape template. For example, these might be the 28 dimensions defining each size template in the SA sizing system.

2   A series of size templates (1 ... s ... p) representing the p size templates in the sizing system. These might be the nine sizes of the SA sizing system (8 to 24).

3   A series of subjects (1 ... j ... m) representing the m subjects in the sample. In the present case, m = 294.

Let

4   S_{ils} be the value of the ith dimension of size template s. For example, this might be the specified Waist girth for SA size 10 (60 cm).

5   M_{ij} be the measurement for the ith dimension on the jth subject. For example, subject 1 may have a Waist girth of 65 cm.

6   W_i be the subjective weighting value for the ith dimension, depending on whether the dimension was too big, or too small (Table 7.1).

Then

7   $|M_{ij} - S_{ils}|$ represents the absolute difference (in cm) between the value of the ith dimension of size template s, and the corresponding measurement on subject j. For the examples above, this value would be 5 (ie $|60 - 65|$).

8   $|M_{ij} - S_{ils}| W_i$ represents the subjectively weighted difference. Note that W_i may vary according to the sign of $(M_{ij} - S_{il})$, and W_i values are means across subjects and are not individualised for each subject. For Waist girth, the weighting is 5.8 for "too small" and 5.6 for "too big". Here $|M_{ij} - S_{ils}| W_i$ would be 5 x 5.8 = 29.0.

9
$$L_{jls} = [\sum_{i=1}^{i=n}(|M_{ij} - S_{ils}| W_i)]/n$$

represents the overall lack of fit for subject j relative to size template s (ie the average of all the lack of fit values for each of the n dimensions).

10
$$minL_j = \min_{s=1}^{s=p} (L_{jls})$$

is the lowest L-value for subject j relative to any of the p size templates in the sizing system. This is the best-fit size.

11
$$L_{tot} = (\sum_{j=1}^{j=m} minL_j)/m$$

is the overall lack of fit for the entire sample of m subjects relative to the sizing system.

Comparisons between the SA sizing system and the measurements of 18–30-year-old Australian women were made using the L-statistic and based on measurements calculated from 3D scans of each participant. A total of 294 Australian women, volunteered to participate in this study, following ethical clearance from the Human Research Ethics committee of the University of South Australia and the Australian Defence Human Research Ethics committee. Subjects were recruited via mass media, from The University of South Australia (Magill and Underdale campuses), Swinburne University and the Australian Institue of Sport.

Twenty-eight measurements, corresponding to the 28 dimensions found in the SA sizing system were calculated from the 3D scan data of each participant. Three-dimensional scans were generated by the *Vitus Smart* laser scanner and *ScanWorX* scanning software (Kaiserslautern, Human Solutions). All measurements were calculated using *CySize* (Perth, Australia, Headus), a measurement extraction program.

Participants were initially guided through a demographic questionnaire, which included questions regarding clothing sizes, such as usual shoe, pants, dress and bra size. Participants changed into tight-fitting underwear for the scan and were provided with a bath robe and slippers, to wear while moving between rooms. If a participant did not have appropriate underwear with them, briefs and sports bras were provided.

A female anthropometrist, accredited at Level 2 by the International Society for the Advancement of Kinanthropometry (ISAK), carried out physical measures, consisting of stretched stature, sitting height, mass and two randomly assigned measures (see Table 7.2), to allow comparisons between scan-calculated and physical measurements. All physical measures were taken according to the ISAK protocol (ISAK, 2001). The anthropometrist then placed physical landmarkers on 22 landmarks, following the CAESAR protocol (Robinette *et al.*, 2002). Participants were asked to wear a swimming cap to reduce missing data on the head and to assist with the placement of the landmarkers. The physical landmarkers used were small balsa wood right-angled triangles. The landmarkers were affixed with double-sided tape, so that the hypotenuse of the triangle was pointing down towards the landmark.

Table 7.2 Physical measurements randomly assigned on testing days.

Girths	Head
	Arm relaxed
	Forearm
	Waist
	Gluteal (hip)
	Calf
	Ankle
Lengths	Acromiale-radiale
	Radiale-stylion
	Midstylion-dactylion
	Trochanterion-tibiale laterale
	Tibiale mediale-sphyrion tibiale
Breadths	Biacromial
	Biiliocristal
	Foot length
	Biepicondylar femur
	Biepicondylar humerus

Once landmarking had been completed, with all 22 landmarks attached, the participant moved through to the *Vitus Smart* laser scanner. When in the scanner, the anthropometrist positioned the participant in a standard scanning posture (Figure 3.1). The participant was asked to maintain that posture and stay as still as possible during the 15 s scan. *ScanWorX* software linked to the scanner produced the 3D scan data. Each scan was examined by an ISAK-accredited Level 2 anthropometrist to ensure that all 22 landmarkers were visible and accurately placed.

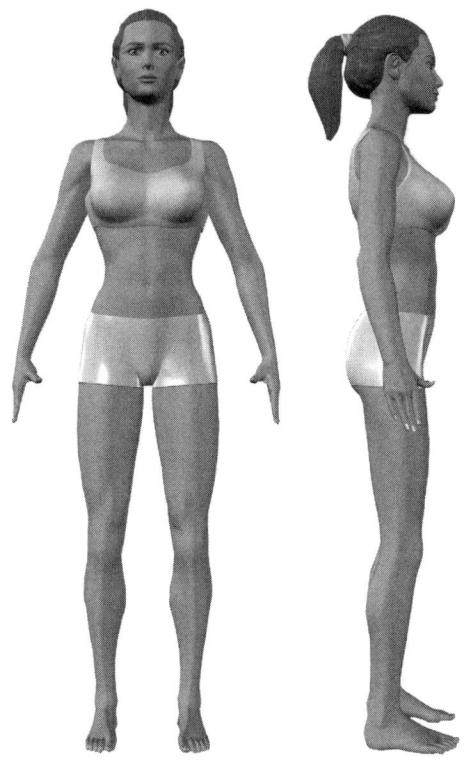

Figure 7.2 The standard scanning posture participants were asked to assume.

Twenty-eight measurements, corresponding to the 28 dimensions found in the SA sizing templates, were calculated from the 3D scan data of each participant (Standards Australia, 1997). Three-dimensional scans were generated by the *Vitus Smart* laser scanner and *ScanWorX* scanning software (Kaiserslautern, Human Solutions). All measurements were calculated using *CySize* (Perth, Australia; Headus), a measurement extraction program. The mean intra-tester error for any measurement was 6.7 mm. When all 28 measurements had been calculated, the measures were exported into custom analytical software *Sizing*, which calculates the L-statistic. High values of L indicated a poor match, while a value of 0 represented a perfect match. An overall best-fit size for each subject was calculated as the size with the lowest L-value.

Distributions of best-fit sizes and reported dress sizes (from the questionnaire) were compared using Cohen's kappa and percentage agreement. Cohen's kappa is a measure of concordance for categorical data and was calculated using SPSS 12.0.1.

4 RESULTS

The mean (± SD) age of the 294 participants was 22.7 (3.2) years, height 168.5 (7.2) cm, mass 65.1 (11.2) kg, and BMI 22.9 (3.6) kg.m^{-2}.

Figure 7.3 shows the z-scores for each SA dimension, in relation to the mean scan-calculated measurement for that dimension. The SA value for each size on each dimension is expressed as a z-score relative to the entire dataset of 294 women. For example, the SA Bust girth (75 cm) is about 2 SD below the overall mean for all women (90.3 ± 7.8 cm). Each line represents one size, with size 8 on the extreme left and size 24 on the extreme right.

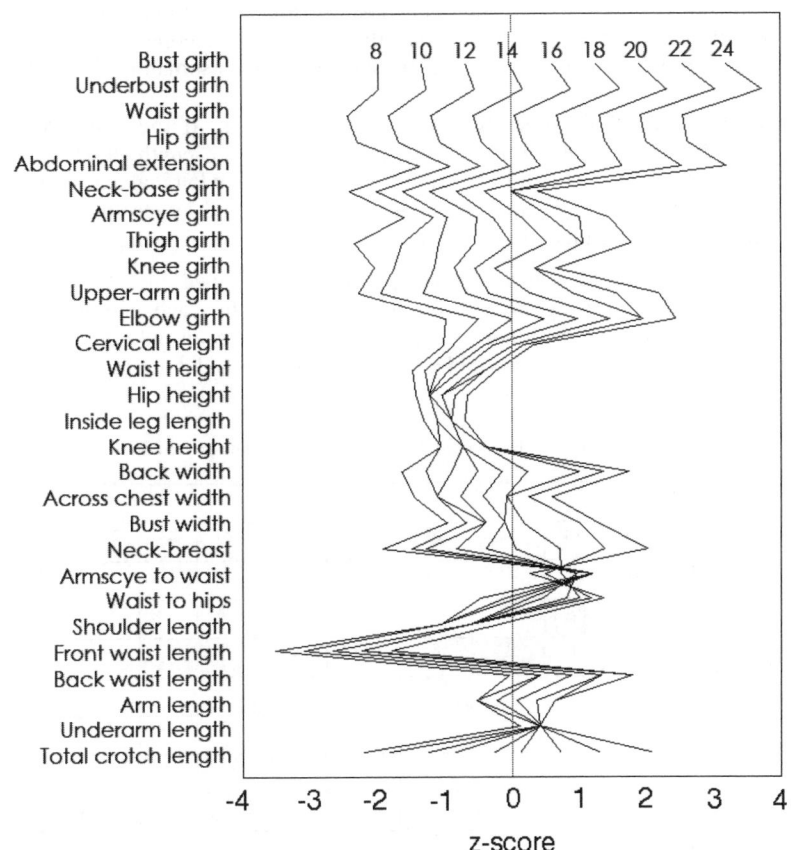

Figure 7.3 Z-scores for each SA dimension compared to mean scan-calculated measurement.

106 Honey and Olds

To interpret this graph, it is important to note the shift, spread and spacing of the lines, as outlined below.

- The *shift* of the lines relative to the mean ($z = 0$). If lines as a whole are generally shifted to the left, the SA dimensions are generally smaller than the scan-calculated measurements. This is obviously the case for Front waist length, which is 2–3 SD below the mean. If the lines are shifted to the right, the SA dimensions are larger than the scan-calculated measurements. Back waist length displays a clear rightward shift, being approximately one SD above the mean.

- The *spread* of the lines should roughly span between –2 and +2 SD, as this would encompass 95% of all women. Hip height for example, as well as showing a leftward shift (templates are too small), also clusters together and therefore does not capture the variability found in our population. Waist girth has a spread which may be slightly too large, overestimating the population variability.

- The *spacing* of the lines represents the distribution. If the lines are evenly spaced, they represent a normal distribution. Shoulder length shows clustering (11 cm for sizes 8–12 and 11.5 cm for size 14–22). Abdominal extension shows uneven spacing, with small spacing at low sizes, and larger spacing at large sizes. Total crotch length, on the other hand, shows even spacing.

Based on the L-statistic, participants were grouped into "best fit" size categories (the size category with the lowest L-value for each individual). Table 7.3 compares optimal size and reported dress size. Data on only 282 of the 294 participants are reported in this table, as some participants did not report their usual dress size.

Table 7.3 Cross-tabulation of optimal size categories and reported dress size.

		Best-fit size										
		8	10	12	14	16	18	20	22	24	all	%
Reported dress size	8	1	6	23	12	3					45	16.0
	10		1	10	60	29	2				102	36.2
	12		1	1	8	49	28	3			90	31.9
	14					4	5	21	2	1	33	11.7
	16							3	1	4	8	2.8
	18								1	2	3	1.1
	20									1	1	0.4
	22										0	0
	24											
	All	1	8	34	80	85	35	27	4	8	282	100
	%	0.4	2.8	12.1	28.4	30.1	12.4	9.6	1.4	2.8	100	

There were significant differences between the distributions of reported dress sizes and best-fit sizes (χ^2 = 539, p < 0.0001). Cohen's kappa was –0.09, and the percentage agreement was only 1%. Reported dress sizes matched best-fit size in only 3 of 282 subjects, and reported dress size was greater than optimal size in only one case. The best-fit size was one size bigger than reported dress size in 10% of women, two sizes bigger in 50% of women, three sizes bigger in 33% of women and four sizes bigger in 5% of women. In one woman, the best-fit size was five sizes bigger than the reported size (24 vs 14). The most common best-fit sizes were 16 (30.1%) and 14 (28.4%), whereas the most common reported dress sizes were 10 (36.2%) and 12 (31.9%). The mean reported dress size was 11.1, while mean best fit size was 15.6, an upward shift of more than two sizes (p < 0.0001). The difference between best-fit size and reported dress size increased as best-fit size increased, from no difference at size 8, to 7.2 at size 24.

Participants were grouped into their best-fit sizes, according to the L-statistic. From these groupings, a best-fit sizing system was developed. This consisted of the mean scan-calculated measurements for each dimension of all the women in each size grouping. These new dimensions became the best-fit sizing system. This is displayed in Table 7.4.

Table 7.4 The best-fit sizing system. All measurements are in cm.

Dimension	8	10	12	14	16	18	20	22	24
C1 Height	157.2	160.5	164.4	167.2	170.7	171.2	171.4	164.9	168.4
C3.1 Bust girth	73.9	78.3	82.5	87.3	90.5	94.8	100.7	105.7	112.3
C3.2 Underbust girth	67.5	68.1	71.7	75.9	79.2	83.0	87.8	91.7	100.2
C3.3 Waist girth	60.3	63.5	67.1	70.8	74.5	79.3	86.0	91.1	100.0
C3.4 Hip girth	81.2	85.9	90.5	94.6	99.6	104.1	108.3	114.4	124.6
C3.5 Abdominal extension	68.8	75.6	79.4	82.7	87.4	92.6	98.8	104.6	116.4
C3.6 Neck-base girth	37.1	38.4	39.1	40.2	41.3	42.3	43.6	43.0	44.4
C3.7 Armscye girth	32.4	35.6	38.1	41.0	42.4	45.2	47.1	50.0	53.9
C3.8 Thigh girth	45.3	50.0	53.7	56.4	59.7	62.8	66.1	69.5	75.5
C3.9 Knee girth	31.0	33.6	35.2	36.3	38.2	39.8	41.0	42.6	47.0
C3.10 Upper-arm girth	22.9	25.9	26.9	28.8	30.4	32.1	34.2	35.5	38.0
C3.11 Elbow girth	19.9	22.3	23.0	24.2	25.3	26.3	27.1	28.0	30.3
C4.1 Cervical height	132.6	135.9	139.6	142.5	146.3	146.5	147.3	141.6	144.9
C4.2 Waist height	96.9	100.2	103.8	106.2	109.3	109.5	109.7	106.0	108.1
C4.3 Hip height	79.3	82.0	84.9	87.0	88.7	88.8	90.0	85.7	86.9
C4.4 Inside leg length	71.8	72.8	75.7	78.0	80.1	80.2	80.4	77.0	77.4
C4.5 Knee height	41.3	42.5	44.6	46.0	47.1	47.3	47.0	44.1	46.0
C5.1 Back width	33.5	31.0	32.5	33.5	35.0	35.8	36.4	36.9	36.6
C5.2 Across chest width	30.0	29.8	31.4	32.4	33.8	34.4	35.2	34.1	38.8
C5.3 Bust width	15.8	16.9	17.3	18.4	18.7	19.3	20.5	22.5	22.0
C5.4 Neck shoulder point to breast point	24.1	24.0	24.8	25.3	26.7	27.2	29.0	29.6	30.8
C5.5 Armscye to waist	19.8	19.5	18.9	19.4	19.3	18.4	18.5	16.0	16.6
C5.6 Waist to hips	18.3	19.1	19.8	20.3	21.6	21.7	20.7	21.5	22.7
C5.7 Shoulder length	11.9	12.4	11.9	11.9	12.4	12.6	12.7	11.1	12.9
C5.8 Front waist length	38.2	38.5	39.6	40.6	41.4	41.9	43.0	42.7	43.9
C5.9 Back waist length	36.5	37.0	37.0	37.7	38.6	38.5	39.1	37.6	39.0
C5.10 Arm length	52.8	56.6	57.2	59.4	60.9	60.8	61.1	60.1	61.1
C5.11 Underarm length	38.5	40.1	41.1	41.9	42.3	41.9	41.2	39.7	39.0
C5.12 Total crotch length	64.9	71.1	74.2	76.0	79.4	81.0	83.1	88.0	91.6

5 DISCUSSION

Current literature suggests that there is a general dissatisfaction among women in relation to clothing sizes and that this dissatisfaction is largely due to poor-fitting garments (Anderson-Connell, Ulrich and Brannon, 2002; Ross, 2003; Simmons and Istook, 2003; Sproles and Geistfeld, 1978; Treleaven, 2004). On comparison of best-fit size (as determined by the L-statistic) and reported dress size, a difference of more than two sizes was noted. The average reported dress size was 11.1, whereas the average best-fit size was 15.6. This would perhaps explain why many women report poor-fitting garments, if they are actually wearing the wrong size. Given that the original SA sizing system ranges in size from 8 to 24, we would expect that the middle size (16) would have been intended to be close to the population average, which matches with our results. With an average reported dress size of 11.1, this suggests substantial sizing drift over time, indicating that the most likely explanation of our findings is vanity sizing.

The average minimum L-value found in our sample (19.9) was significantly different from zero ($t = 56.6$, $p < 0.0001$). This indicates a significant difference between the SA sizing system and the 3D shapes of real Australian women. To devise a new and better sizing system, the L-statistic could be used to review the current sizing system and new dimensions could be put forward. Alternatively, a completely new sizing system could be developed, with new dimensions and different clothing size categories.

The current system could be reviewed by capturing a larger data set, with a broader scope and creating a best-fit sizing system, using the L-statistic. The best-fit sizing system could then be substituted for the current system. This would fit the women of the larger data set, as development would be based upon their measurements.

To follow on from merely substituting a best-fit sizing system for the SA dimensions, a recursive algorithm could be developed, whereby L-values for all participants are continuously calculated, to refine the best-fit sizing system. Each time L-values were calculated for the sample, each participant would be assigned to a best-fit category and a new best-fit sizing system would be developed based on the average dimensions from the latest categorisation. This would continue until the overall L-value failed to show significant reductions. By recursively modifying the optimal sizing system, one could refine the system to better fit the population of interest.

If a new sizing system is developed, not only should the size of the dimensions be reviewed, but also the definitions of the dimensions. Many of the dimensions in the SA document were poorly defined. A new sizing system should incorporate clear definitions that allow equivalent physical measures to be taken, to allow continual revision of the system using the L-statistic, in a more accurate and efficient manner.

Future work into the L-statistic and new clothing sizes should be in close conjunction with 3D body scanners. The data generated by a 3D scan allow the production of a whole new dimensional set. This means that instead of simple lengths and girths making up the dimensions of a sizing system, new dimensions such as angles (neck-shoulder angle, and waist angle) and cross-sectional shapes (waist ellipse) can be used to further refine clothing sizes to fit real bodies.

"Mass customization" has been frequently talked about in relation to 3D scanning and clothing production. Mass customisation is the production of perfect-fitting garments. An infinite number of measurements (including angles and cross-sectional shapes) can, in principle, be calculated from 3D scan data, allowing the generation of a pattern, and thereby garment, fitted to the individual. Three-dimensional scan data could also exploit new garment production technologies. The production of seamless garments has recently been introduced. By combining this technology with 3D scan data, circular looms could produce seamless garments direct from the 3D scans. This technology could be applied to garments such as wetsuits, burns suits, swimwear, ski and cycling suits, and thermal underwear, where a close fit is critical.

The use of 3D body scanners and the L-statistic has potential for use in revising the current sizing system, for continuous revision of future sizing systems and for the development of new clothing technologies. To allow this, changes need to be made to current 3D scanning procedures.

A standardised body scanning protocol should be developed for worldwide use. Several studies have been conducted in the UK, the US and now here in Australia. A recent investigation has found that measurements taken using two different scanner-software combinations are comparable (Daniell, 2005). Therefore the data from different projects can theoretically be compared. However, different landmarks and different measurements are used. The International Organization for Standardization (ISO) is currently developing an anthropometric standard that specifies methodologies and protocols for use with 3D scanning systems. However, this standard should also include a set of landmarks and measurement definitions, specific to 3D body scanning.

To further minimise the time required for large scale anthropometric studies using 3D scanners, automatic landmark recognition and automatic measurements need to be further developed to a point where they are as accurate as physical measurements. Without this, anthropometric surveys are too expensive, in terms of time required by trained operators, to be viable for continuous revision of the sizing systems.

The L-statistic was developed for this study to determine the mismatch between the shapes assumed by the SA sizing system and those found in a population of young Australian women. Overall, large differences exist and provide evidence for promotion of the development of a new sizing system. The development of a new sizing system, or new garment production technologies should be done using the L-statistic and in close association with the development of 3D body scanners.

6 REFERENCES

AADBase, 1995, *Australian Anthropometric Database* (Adelaide: School of Health Sciences, The University of South Australia).

Anderson-Connell, L.J., Ulrich, P.V. and Brannon, E.L., 2002, A consumer-driven model for mass customisation in the apparel market. *Journal of Fashion Marketing and Management*, **6**(3), 240–258.

Biederman, M., 2003, A bulge in misses 8? Digital scanners resize America. *The New York Times*, 27 November, p. 8.

Cole, T.J., 2003, The secular trend in human physical growth: A biological view. *Economics and Human Biology*, **1**(2), 161–168.

Daniell, N.D., 2005, *A Comparison of the Accuracy of the Vitus Smart and Hamamatsu Body Line 3D Whole-body Scanners*, unpublished thesis, University of South Australia.

International Society for the Advancement of Kinanthropometry (ISAK) 2001, *International Standards for Anthropometric Assessment* (Underdale, SA, Australia).

Lahti-Koski, M., Pietinen, P., Mannisto, S., and Vartiainen, E., 2000, Trends in waist-to-hip ratio and its determinants in adults in Finland from 1987 to 1997. *American Journal of Clinical Nutrition*, **72**, 1436–1444.

Lang, S.S., 1995, Designers size clothes to reflect different body proportions. *Human Ecology*, **23**(2), 3.

Pontoni, F., 2004, First national clothing sizes forum. *Standards Australia Press Release*, February 2004.

Robinette, K.M., Blackwell, S. and Daanen, H., 2002, *Civilian and European Surface Anthropometry Resource (CAESAR) Final Report* (Philadelphia: SAE International).

Ross, T., 2003, A fitting solution. *Apparel Magazine*, **44**(12), 42–43.

Simmons, K.P. and Istook, C.L., 2003, Body measurement techniques, comparing 3D body-scanning and anthropometric methods for apparel applications. *Journal of Fashion Marketing and Management*, **7**(3), 306–332.

Sproles, G.B. and Geistfeld, L.V., 1978, Issues in analysing consumer satisfaction/dissatisfaction with clothing and textiles. *Advances in Consumer Research*, **5**(1), 383–391.

Standards Australia, 1997, *Australian Standard: Size Coding System for Women's Clothing – Underwear, Outerwear and Foundation Garments* (NSW, Australia: Standards Australia).

Thomas, C., 2004, The shape we're in. *New Scientist*, **184**(2471), 42–43.

Treleaven, P., 2004, Sizing us up. *IEEE Spectrum*, **4**(41), 29–31.

Ulijaszek, S.J., 2001, Increasing body size among adult Cook Islanders between 1966 and 1996. *Annals of Human Biology*, **28**(4), 363–373.

Usher, R., 1996, A tall story for our time. *Time Australia* (46), 126–132.

CHAPTER EIGHT

Body composition in female sports participants with particular reference to bone density

J. Wallace, T. Donovan, K. George and T. Reilly

Research Institute for Sport and Exercise Sciences, Liverpool John Moores University, Henry Cotton Campus, 15-21 Webster Street, Liverpool, L3 2ET

1 INTRODUCTION

Osteoporosis is a disease characterised by low bone mass and the microarchitectural deterioration of bone tissue leading to an increase in bone fragility and the consequent risk of fracture. Osteoporosis affects 8 to 10 million people in the United States and leads to approximately 1.5 million fractures annually (Melton III, 1995); it is one of the most debilitating and costly diseases of modern society, placing a huge strain on the National Health Service and its resources. The current major public health strategy is to emphasise preventing rather than treating osteoporosis. The aims are to enhance peak bone mass and reduce bone loss during early to late adulthood (International Federation of Sports Medicine, 2000).

The relationship between mechanical loading and the development and maintenance of peak bone mass is well accepted and has been defined as Wolff's law: "Every change in form and function of a bone, or in its function alone, is followed by certain definite changes in its internal architecture and equally definite secondary alteration in its mathematical laws" (Wolff, 1986). Bone adapts its external shape and internal structure to provide maximum strength with minimum mass in response to mechanical forces that it is required to support. Reductions in bone loading due to paralysis (Bauman *et al.*, 1999), space flight (Vico *et al.*, 2000) or immobilisation (Zerwekh *et al.*, 1998) result in a loss of bone mass, whereas increased mechanical loading due to exercise can result in bone mass accrual and impede or prevent degeneration due to aging. These adaptations of bone tissue are regulated by genetic and systematic factors, mechanical forces, hormones and growth factors. The specific role of physical activity and sports in the maintenance or enhancement of bone mass architecture remains elusive, despite considerable research attention.

Quantitative dose-response studies are lacking, making it not yet possible to describe in detail an exercise programme that will optimise peak bone mass. However, some consensus among recommendations for the type, intensity, frequency and duration of exercise that best augment bone mineral accrual has been achieved (Kohrt *et al.*, 2004).

The primary variables associated with mechanical loading on bone are the magnitude of the strain induced and the rate at which the strain is imposed. The strain magnitude is positively related to the osteogenic response, but increasing evidence suggests that strain rate is also an important factor (Kohrt *et al.*, 2004). Increasing strain rate, while holding loading frequency and peak strain magnitude constant, was found to be a positive determinant of change in bone mass (Burr, 2002).

Load intensity (strain magnitude) is the most important factor for promoting osteogenic responses and peak bone mass (Rubin and Lanyon, 1985). Mechanical forces have to produce unique, variable and dynamic stress to bone in order to have an osteogenic effect, while static loading of bone does not cause the osteogenic adaptive response that occurs with dynamic loading (Burr, 2002). Only skeletal sites exposed to a change in daily loading forces undergo adaptation and therefore the effect is site specific. The type of forces applied to bone also produces different osteogenic effects; compression forces are better than sheer and torsional forces and unevenly distributed strains at a site show a potential to increase bone mineral content at that specific site.

During adulthood, it is recommended that weight-bearing endurance activities should be performed 3–5 times per week combined with resistance exercise 2–3 times per week, for a recommended duration of 30–60 minutes a day (Kohrt *et al.*, 2004). Rubin and Lanyon (1985) demonstrated that only a few external-loading cycles (e.g. 36 per day) of a relatively high magnitude were necessary to optimise the bone formation response, increasing the number of loading cycles beyond this had no additional effect.

Body composition is also a very important consideration for athletes. Excessive adipose tissue acts as a dead weight in activities where the body mass must be repeatedly lifted against gravity during locomotion and jumping (Reilly, 1996), slowing the performance down, and increasing the energy demands of the activity. In contrast, lean mass contributes to power production during high-intensity activity and provides a greater absolute size and strength for resistance against high dynamic and static loads.

The purpose of this cross-sectional study was to compare bone mineral measurements and body composition among female participants at club level in two different sports (Rugby Union and Rowing) against age-matched sedentary controls. These unique observations would help determine which sports participants achieve the greater peak bone mass for best protection against osteoporosis. A secondary aim was to examine the difference in body composition and bone density between playing positions in Rugby Union football.

2 METHODS

2.1 Subjects

Eighty-five female subjects aged 18–43 years, who participated in sport at club level or above, took part in the study. The sample included rugby union players (N = 27) (forwards [N = 15] and backs [N = 12]) rowers (N = 28) and controls (N = 30). The mean ± SD characteristics were: Age: 22.67 ± 5.01 years; Mass: 66.39 ± 9.60 kg; Height: 1.67 ± 0.09 m. All subjects received a full written and verbal explanation of the nature of the study before providing informed written consent. Subjects with metal implants, prostheses or metal jewellery that could not be removed were excluded from the study as were those with any known bone disease.

2.2 Body composition measurement

Subjects arrived at the laboratory in the morning, after fasting for 3 hours. Participants wore light clothing (for example t-shirt and shorts) without zips, buttons, under-wiring or any other metal and removed all jewellery for the scanning. Subjects' age, height (m), and body mass (kg) on beam-balance scales were measured. Whole-body mass, fat mass, lean mass, percent body fat (%BF), bone mineral content (BMC) and bone mineral density (BMD) were then measured according to standard operating procedures using a fan beam dual-energy X-ray absorptiometry scanner (DEXA; Hologic QDR series, Delphi A, Bedford, Massachusetts). Scans were analysed using system Hologic QDR Software for Windows version 11.2 (© 1986–2001 Hologic Inc.) according to standard analysis protocols. All scans were analysed by the same investigator. All procedures received approval from Liverpool John Moores University's Human Ethic's Committee.

The coefficients of variation (CV) (Equation 8.1) of repeated total whole-body DEXA measurements in our laboratory ranged between 0.71% and 1.25% for total mass, fat mass, lean mass, percent body fat, bone mineral content and bone mineral density. Values for technical error of measurement (TEM) (Equation 8.2) of repeated whole-body measures in our laboratory for BMD, total mass, fat mass, lean mass and fat (%) were 0.01, 0.24, 0.20, 0.30 and 0.34 respectively. The 95% Limits of agreement ranged from 0.00 ± 0.02 for whole-body BMD, up to 0.07 ± 0.55 for whole-body fat mass.

$$CV = \frac{\left(\frac{SD_{diff}}{\sqrt{2}}\right)}{\overline{X}} x100 \qquad (8.1)$$

Where SD_{diff} is the standard deviation of the total differences between the measures; and $\overline{X}$ is the mean of all values

$$TEM = \sqrt{\frac{\sum d^2}{2n}}$$ (8.2)

Where d is the difference between each pair of measures; and n is the sample number.

2.3 Statistical analysis

Results are expressed as mean ± SD where relevant. Statistical analysis was carried out using SPSS version 12. A one-way ANOVA was used to determine if significant ($P < 0.05$, $P < 0.01$) differences existed among the sports groups and between the positional roles for participant descriptives, body composition and BMD variables. When significance was found, a Tukey post-hoc test was performed.

3 RESULTS

3.1 Body composition relative to sport

The descriptive data for each sports group are presented in Table 8.1. The rugby players have been further broken down into their positional roles of forwards and backs. The sedentary controls and rugby players were of a similar age irrespective of playing position, yet the rowers were significantly older. There was no significant difference in height between the sport groups or between playing position. Only the rugby players had a significantly greater mass than the sedentary controls ($P < 0.05$) and there were no differences between the rowers and either the rugby players or controls. However, when the rugby players were analysed with respect to their playing position, the forwards were significantly heavier in body mass than the sedentary group, the rowers and the rugby back players. The rugby players who played in back positions were not significantly different in total mass from either the rowers or the control group.

Table 8.1 Descriptive data of subjects, sorted by group (mean ± SD).

Sport (n)	Age (Years)	Height (m)	Mass (kg)
Rowers (28)	27.4 ± 5.9[acde]	1.69 ± 0.06	65.72 ± 7.89[d]
Rugby Union (27)	20.6 ± 1.4[b]	1.68 ± 0.06	71.33 ± 11.59[a]
– Forwards (15)	20.3 ± 1.4[b]	1.70 ± 0.06	77.81 ± 10.80[abe]
– Backs (12)	20.9 ± 1.4[b]	1.66 ± 0.06	63.22 ± 6.33[d]
Sedentary (27)	20.1 ± 2.5[b]	1.63 ± 0.12	62.53 ± 6.90[cd]

[a] is significantly different from sedentary   [b] is significantly different from rowers
[c] is significantly different from Rugby Union   [d] is significantly different from forwards
[e] is significantly different from backs

The mean total mass of the rowers and the rugby players as measured by the DEXA method were both greater than the sedentary controls, although only the total mass of the rugby union players was significantly greater (P < 0.001). The significantly greater total mass of the rugby players in comparison to the sedentary controls was apparent in both the forward (P < 0.001) and back (P < 0.001) playing positions. There was no significant difference observed in total mass between the two sports (P > 0.05).

The rugby union players had a significantly (P < 0.05) larger fat mass than the control subjects (22.54 ± 8.09 kg and 17.17 ± 4.52 kg, respectively). However, when the rugby players were split into their playing positions it was only the forwards who had a significantly greater (26.55 ± 8.39 kg; P < 0.001) fat mass than the controls, whereas those playing as backs had a comparable fat mass to the sedentary controls (17.52 ± 3.98 kg). The rowers had a slightly lower fat mass (15.74 ± 4.56 kg; P > 0.05) in comparison to the controls. Rowers also had a significantly lower (P < 0.001) fat mass than the rugby players as a whole, and with those playing in the forwards (0.001). No difference was noted between the rowers and those who played as a back in rugby (P > 0.05).

Both rugby union players and the rowers had a significantly greater lean mass than the sedentary controls (controls: 43.14 ± 4.10 kg; rugby: 48.17 ± 6.85 kg, P < 0.01; rowers; 47.83 ± 4.44 kg, P < 0.05), although there was no difference between the two sports (P > 0.05). Those in forward playing positions in rugby showed an average 8 kg, larger lean mass (51.26 ± 6.80 kg, P < 0.001) than the control group; there was no difference between the controls (43.14 ± 4.10 kg) and the backs (44.31 ± 4.78 kg) in their lean mass.

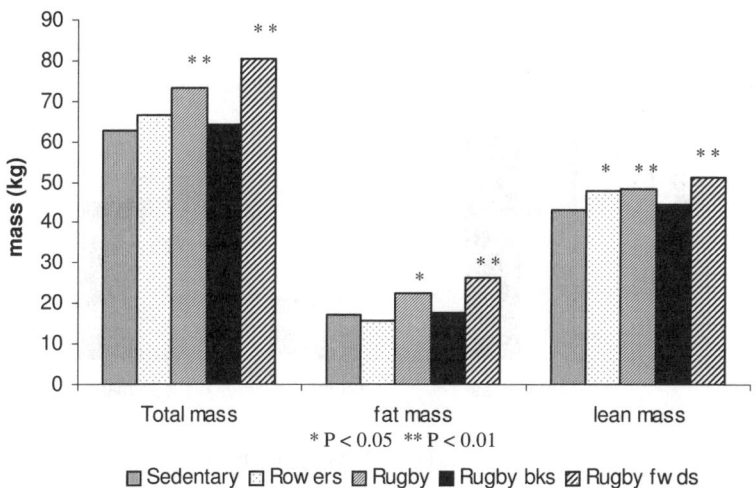

Figure 8.1 Body composition of female rugby union players, rowers and sedentary controls.

The rowers demonstrated a lower percent body fat (23.98 ± 4.11%) and the rugby players showed a greater percent body fat (30.11 ± 6.11%) and a greater range of values compared to the sedentary controls (27.18 ± 5.06%), although neither of these differences were significant (P > 0.05). There was a significant difference between the rowers and the rugby union players in total percent body fat values (P < 0.001), which equated to an average difference of 6 % between the two sports. When adjusted for playing position, the difference in percent body fat between the rugby players and the rowers was only observed in the forwards (32.47 ± 6.28%, P < 0.001). The body fat percent values of the rugby union backs (27.18 ± 4.59%) were the same as for the sedentary controls (27.18 ± 5.06%).

3.2 Body composition of different playing positions

As noted in the previous section, there were many differences in body composition between the playing positions in rugby union and the sedentary controls. There was also a large difference between the playing positions alone (Figure 8.2). The forward players had a significantly greater total body mass (80.39 ± 13.28 kg, P < 0.001) than the backs (64.13 ± 6.97 kg), attributed to a significantly greater fat mass (P < 0.01) and also a greater lean mass (P < 0.05). The forwards were on average 16 kg heavier than the backs, with almost 10 kg of this being fat mass. Although the forward players had both an increased fat mass and lean mass (26.55 ± 8.39 kg and 51.26 ± 6.80 kg) compared to the backs (17.52 ± 3.98 kg; 44.31 ± 4.78 kg respectively), they still had a significantly higher percent body fat as well (forwards: 32.47 ± 6.38%; backs: 27.18 ± 4.59%). This finding is due to the fact that the difference in total fat mass was greater than the difference in lean mass between the forwards and backs.

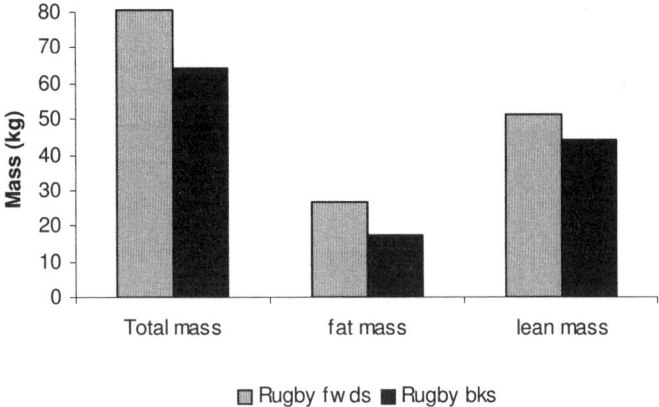

Figure 8.2 Body composition comparisons between rugby union forwards (fwds) and backs (bks).

3.3 Bone density of different sports

Figure 8.3 shows that both rugby union players (2452 ± 446 g) and rowers (2467 ± 342 g) had a higher bone mineral content (BMC) than the sedentary controls (2218 ± 305 g). These differences did not reach the required level of statistical significance (P = 0.103 and P = 0.150 respectively). Only those rugby players competing in forward positions were found to have a significantly greater bone mineral content (2579 ± 464 g) in comparison with the sedentary controls. The backs displayed the smallest difference in bone mineral content (2294 ± 384 g) in comparison to the controls.

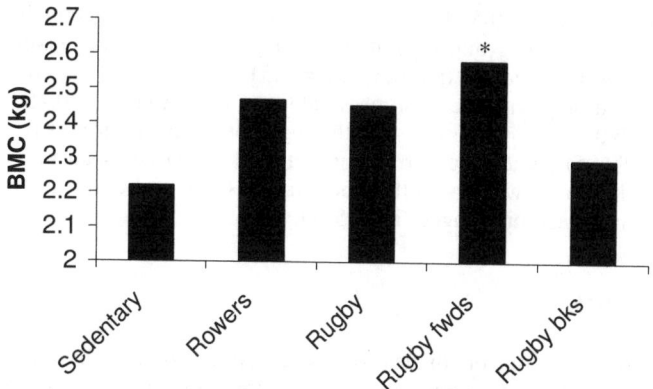

Figure 8.3 Comparison of bone mineral content (BMC) between sports.

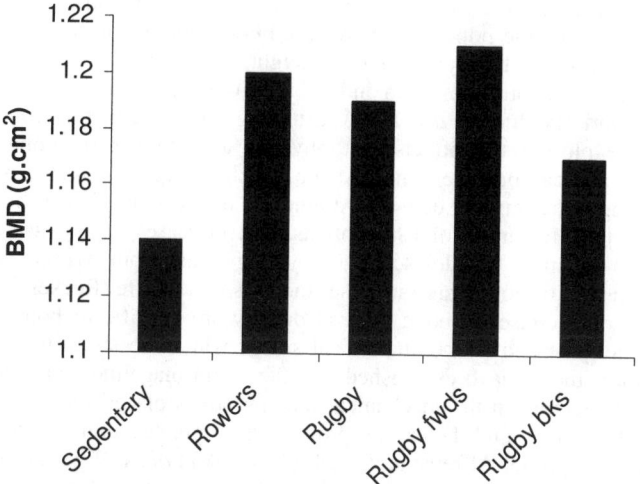

Figure 8.4 Comparison of bone mineral density (BMD) between sports.

The bone mineral density (BMD) did not significantly (P > 0.05) vary between groups: sedentary controls (1.14 ± 0.10 g.cm^2), rugby union players (1.19 ± 0.12 g.cm^2) or the rowers (1.20 ± 0.09 g.cm^2). However, when the two sports groups were pooled together, there was a significant difference between them and the control group (P < 0.05). There was also no significant difference between the playing positions in the rugby group. Although there were no significant differences between the type of sports participation, Figure 8.4 demonstrates that both rugby players and rowers showed higher values for bone mineral density than the control group. The largest differences were observed between the rugby forwards (1.21 ± 0.13 g.cm^2) and rowers (1.20 ± 0.09 g.cm^2) compared with the sedentary control group (1.14 ± 0.10 g.cm^2). Those playing in the back positions (1.17 ± 0.10 g.cm^2) in rugby displayed the smallest difference in comparison to the controls.

Bone mineral density was correlated with the total mass (r = 0.335; P < 0.01) and the lean mass (r = 0.559; P < 0.01) of all the subjects collectively. The BMD values for each sports group separately (rugby, rowers, sedentary) were correlated with their lean mass values (r = 0.534; r = 0.518; r = 0.480; P < 0.01), but BMD was correlated only with the total mass values of the rowers (r = 0.450; P < 0.05), and not with the total mass values for the rugby or sedentary groups (r = 0.272; r = 0.199; P > 0.05). When the correlation between BMD and lean mass for the all subjects collectively was controlled for total mass by partial correlation analysis, the r-value remained unchanged (r = 0.551).

4 DISCUSSION

A population of club-level athletes was examined in this study in order to compare bone densities and body composition of an impact-loading sport (rugby union) to that of an active-loading sport (rowing). Different sport groups are unique because the participants have been trained using similar bone-loading situations during their training, which is specific to the demands of their sports and varies from one sport to another. The athletic population may be a biased population owing to a genetic make-up of physical characteristics (e.g. height, somatotype) that favours athletic success and may contribute to an individual's predisposition to participate in a particular sport (Fehling *et al.*, 1995). Studying various athletic groups offers a method of exploring the effects that physical activity has on bone and body composition and can provide a model from which to explain possible mechanisms for enhancing bone mineral density. Attainment of an optimal peak bone mass is essential in the prevention of osteoporosis, and increases in physical activity or participation in sport in adolescent and young adult women may significantly reduce the risk of osteoporosis and bone fractures in later life (Recker *et al.*, 1992). The effects of exercise on bone mineral density are specific to both the type of activity undergone and to the anatomical site at which the strain load is applied. Several other studies have established that the strain magnitude and rate of strain are important factors in the mechanical loading stress placed on the body during high-impact activity and hence in promoting a greater bone mineral density (Heinonen *et al.*, 1995; Alfredson *et al.*, 1997; Dook *et al.*, 1997; Matsumoto et al., 1997). Thus sports that assist the attainment of an optimal peak bone mineral density should be identified and promoted.

The main finding of this current study was that the group of female athletes who were involved in the sport with the highest impact loading on the skeletal system displayed a greater bone mineral content and bone mineral density than the athletes who participated in only active-loading sports, or the sedentary group. The Rugby players with roles in forward positions were the only group in this study to have a significantly greater bone mineral content than the sedentary control group; this finding supports the view that the higher the strain rate and magnitude of the impact experienced in a sport, the greater the bone mineral content of an individual is liable to be. It was also observed that both the rowers and Rugby players as a combined sports group displayed a significantly greater ($P < 0.05$) BMC than the sedentary controls. Positionally only the forwards were found to have a significant difference in BMC in comparison to the control group. The bone mineral densities of the combined sports groups in this study were significantly different from those of the sedentary controls. This finding agrees with the notion that participation in sport and training increases bone mineral density of the players (Rubin and Lanyon, 1985). The rowers but not the rugby players proved to have a significantly ($P < 0.05$) greater BMD than the control group, although the sport of rugby is considered to be active loading rather than an impact loading sport. When the rugby players were divided into their positional roles, forwards were also found to have significantly greater bone mineral density than the sedentary control group, whilst the backs displayed no significant difference. The difference in BMD between the forwards and backs in rugby union is probably the main reason why the rugby players as a whole group showed no significant difference in their BMD compared to the sedentary control group.

Studies have shown that bone mass in athletic populations can still be increasing well beyond their early twenties, whilst peak bone mass is often achieved earlier (early twenties) in sedentary individuals (Lu *et al.*, 1994). The rugby union players were still young (20.6 ± 1.4 years) and might not yet have reached their peak bone mass. In comparison the rowing group had an older average age of 27.4 ± 5.9 years and their peak bone mass had probably been attained. The individuals in this study had not been involved in their sport for that many years (average of 3.7 years), and some were only playing at a club level. Longer participation in the sport and a higher level of participation are likely to have an even more pronounced effect on the bone mineral content and bone mineral density, by prolonging bone mass accrual.

Mean bone mineral density and bone mineral content were greater in both the rugby union players and the rowers (though not always significantly). Rugby union has been associated with increased BMD in males (Bell *et al.*, 2003) and females (Egan *et al.*, 2006). This game is becoming increasingly popular among females and could be promoted in teenagers as a beneficial sport for promoting bone health and for the long-term prevention of osteoporosis after the menopause. The impact nature of this contact sport is likely to be an important factor in the high BMD values reported; therefore it may be more skeletally beneficial for players engaged in the forward positions. Rowing generates a large strain magnitude (force) specific to the vertebral column and through the legs, hips and shoulders and these loading forces have been linked with bone mass accrual. Lariviere *et al.* (2003)

concluded that rowing was associated with increased BMD values in both novice and experienced female rowers and the results support the theory that force magnitude (more than load repetition) is a key variable promoting osteogenesis.

A major component of an athlete's body is its skeletal muscle, which generates force for the situations in the game where high impacts are transmitted and received. Thus lean mass, possibly through the forces applied by the skeletal muscle pulling on bone during movement is an important predictor of bone mineral density (Takada *et al.*, 1997). The lean mass values in this current study were correlated with BMD within each group and across all the groups collectively. Lean mass (skeletal muscle) growth is associated with athletic training and strength training, so it is expected that the sports groups would have a greater lean mass and also an associated increase in BMD in comparison to non-active females. The rowers and the rugby union players both had a significantly greater (P < 0.05) lean mass compared to the sedentary controls. Post hoc analysis demonstrated that it was only those in forward playing positions in rugby who had a significantly greater lean mass, the backs and sedentary controls being comparable for mean lean mass. The training of rugby players, especially the forwards, involves regimens of aerobic and anaerobic activities alongside intense strength and power training, in order to increase muscle mass and strength (Luger and Pook, 2004). In rugby, forwards are particularly involved in scrums, rucks and mauls, where there is vigorous use of the legs and arms and in which an increased lean mass is beneficial. The current findings support this idea that forwards in rugby have a greater absolute lean mass then both their comrades playing in back positional roles and sedentary individuals.

Rowing is a power event, muscle strength and endurance being more important than muscle bulk. The rowers in the current study demonstrated a greater lean mass than control subjects. The rowers whose goal it is to maximise absolute power output by increasing lean tissue, do not necessarily require a heavy build as long as they can achieve a large power output. There is a fine balance for rowers between increasing muscle mass (and therefore power output) and any excess weight in the boat that increases the resistance in the water, which is magnified particularly in the lightweight rowing category. Any excess weight carried by the rower in the form of adipose tissue can be detrimental to performance. Therefore, it would be advantageous for the rowers to lose body fat mass and increase lean tissue mass (Morris and Payne, 1996). Results of the present study demonstrate that rowers are achieving these goals, but there is still room for the female rowers to improve.

The female rowers in this present study had a mean percent body fat of 24.0 ± 4.1%. Morris and Payne (1996) found similar results using skinfold callipers (23.4 %) in their study of female rowers post-season, a measurement time period similar to that in this present study. Morris and Payne also demonstrated significant changes in percent body fat over the duration of the season, with a lowest value (19.68%) during the competitive season.

The rugby union players had a very large total fat mass (22.54 ± 8.09 kg) equating to a mean body fat percent of 30.11 ± 6.11% and those playing in the forward positions were found to have a mean percent body fat of 32.5 ± 6.3%. Due to the high impact nature of the game it is often (more traditionally) perceived that a higher level of body fat is necessary to safeguard against injuries. However, the values observed in this current study are a concern; over 44% of the rugby players had a body fat percentage of 30+%, and almost one in five could be classified as obese, with a body fat percentage over 35% (WHO, 2003). Only a quarter of the rugby players were found to have a body fat of < 25%, the recommended values for an athletic female population (Lohman *et al.*, 1997). These high values of fat mass within the rugby players are not only detrimental to their performance, but could also lead to serious health issues. It should be noted that the fat mass and percent body fat values of those playing in the back positions, were comparable to those of the sedentary controls. It is important to remember that the majority of the participants in this present study were club level players and these observations would probably not be seen in elite players.

5 CONCLUSIONS

In this study, BMD was higher in the females rowers and rugby union (forwards only) players compared with sedentary controls. This difference emphasizes the potential of sport participation in young adults for optimizing peak bone mass and reducing the risk of osteoporotic fractures in the long term. Rowing and rugby union (forwards) appear to be the most beneficial in maximizing lean mass. However, there was an undesirable amount of fat mass and %BF associated with the rugby players in the forward positions, likely due to lifestyle factors still associated with the female game, moderate training programmes undertaken (as rugby union for females is still non-professional) and self-selection can also not be ruled out. These high fat values would probably not apply to high-performance rugby players.

Whilst each sport may have an attraction for specific skeletal characteristics, the differences observed between rugby union and rowing and between rugby union forwards and backs suggest that the type of sport is important in optimizing peak bone mass and defining body composition.

6 REFERENCES

Alfredson, H., Nordstrom, P. and Lorentzon, R., 1997, Aerobic work-out and bone mass in females. *Scandinavian Journal of Medicine and Science of Sports*, **7**, 336–341.

Bauman, W.A., Spungen, A.M. and Wang, J., 1999, Continuous loss of bone during chronic immobilization: A monozygotic twin study. *Osteoporosis International*, **10**, 123–127.

Bell, W., Evans, W.D., Cobner, D.M. and Eston, R., 2003, Whole-body and regional bone mineral density and bone mineral mass in Rugby Union players: Comparison of forwards, backs and controls. In *Kinanthropometry VIII – Proceedings of the 8th International Society for the Advancement of Kinanthropometry (ISAK)*, edited by Reilly, T. and Marfell-Jones, M. (London: Routledge), pp. 143–150.

Burr, D.B., Robling, A.G. and Turner, C.H., 2002, Effects of biomechanical stress on bones in animals. *Bone*, **30**, 781–786.

Dook, J., James, C., Henderson, N. and Price, R., 1997, Exercise and bone mineral density in mature female athletes. *Medicine and Science in Sports and Exercise*, **29**, 291–296.

Egan, E., Reilly, T., Giacomoni, M., Redmond, L. and Turner, C., 2006, Bone mineral density among female sports participants. *Bone*, **38**, 227–233.

Fehling, P.C., Alekel, L., Clasey, J., Rector, A. and Stillman, R.J., 1995, Comparison of bone mineral densities among females in impact loading and active loading sports. *Bone*, **17**, 205–210.

Heinonen, A., Oja, P., Kannus, P., Sievanen, H., Haapasalo, H. and Manttari, A. and Vuori, I., 1995, Bone mineral density in female athletes representing sports with different loading characteristics of the skeleton. *Bone*, **17**, 197–203.

International Federation of Sports Medicine, 2000, FIMS-Positional Statement; The female athlete triad. *FIMS*.

Kohrt, W.M., Bloomfield, S.A., Little, K., Nelson, M. and Yingling, V.R., 2004, Physical activity and bone health. ACSM positional stand. *Medicine and Science in Sports and Exercise*, **37**, 1985–1996.

Lariviere, J.A., Robinson, T.L. and Snow, C.M., 2003, Spine bone mineral density increases in experienced but not novice collegiate female rowers. *Medicine and Science in Sports and Exercise*, **35**, 1740–1744.

Lohman, T.G., Houtkooper, L.B. and Going, S.B., 1997, Body composition assessment: Body fat standards and methods in the field of exercise and sports medicine. *ACSM Health Fitness Journal*, **1**, 30–35.

Lu, P., Brody, J., Ogle, G., Morely, K., Humphries, I., Allen, J., Howman-Giles, R., Sillence, D. and Cowell, C., 1994, Bone mineral density of total body, spine, and fermoral neck in children and young adults: A cross-sectional and longitudinal study. *Journal of Bone Mineral Research,* **9**, 1451–1458.

Luger, D. and Pook, P., 2004, *Complete Conditioning for Rugby* (Champaigne, IL: Human Kinetics).

Morris, F.L. and Payne, W, R., 1996, Seasonal variations in the body composition of lightweight rowers. *British Journal of Sports Medicine*, **30**, 301–304.

Matsumoto, T., Nakagawa, S., Nishida, S. and Hirota, R., 1997, Bone density and bone metabolic markers in active collegiate athletes: Findings in long-distance runners, judoists, and swimmers. *International Journal of Sports Medicine*, **18**, 408–412.

Melton III, L.J., 1995, How many women have osteoporosis now? *Journal of Bone Mineral Research*, **10**, 175–177.

Recker, R.R., Davies, K.M., Hinders, S.M., Heaney, R.P., Stegman, M.R. and Kimmel, D.B., 1992, Bone gain in young adult females. *Journal of the American Medical Association*, **268**, 2403–2408.

Reilly, T., 1996, Fitness assessment. In *Science and Soccer*, edited by Reilly, T. (London: E. & F. Spon), 25–50.

Rubin, C.T. and Lanyon, L.E., 1985, Regulation of bone mass by mechanical strain magnitude. *Calcified Tissue International*, **37**, 411–417.

Takada, H., Washino, K. and Iwata, H., 1997, Risk factors for low bone mineral density among females: The effects of lean body mass. *Preventative Medicine*, **26**, 633–638.

Vico, L., Collet, P. and Guignandon, A., 2000, Effects of long-term microgravity exposure on cancellous and cortical weight-bearing bones of cosmonauts. *Lancet*, **355**, 1607–1611.

WHO, 2003, Global strategy on diet, physical activity, and health. Geneva, World Health Organisation.

Wolff, J., 1986, *The law of bone remodelling [Das Gesetz der Transformation der Knochen]* (Berlin: Springer-Verlag).

Zerwekh, J.E., Ruml, L.A., Gottschalk, F. and Pak, C.Y.C., 1998, The effects of twelve weeks of bed rest on bone histology, biochemical markers of bone turnover, and calcium homeostasis in eleven normal subjects. *Journal of Bone Mineral Research*, **13**, 1594–1601.

CHAPTER NINE

Body composition changes in professional soccer players in the off-season

J. Wallace, E. Egan, J. Lawlor, K. George and T. Reilly

Research Institute for Sport and Exercise Sciences
Liverpool John Moores University, Henry Cotton Campus
15-21 Webster Street, Liverpool, L3 2ET.

1 INTRODUCTION

In recent years many soccer-specific studies have been published, yet very few focus on repeated measurements in professional players. This is perhaps because soccer players are not frequent visitors to the laboratory during the busy season, except for health or contractual reasons at the beginning of a season (Casajús, 2001). In sports such as soccer where the competitive programme occupies nine months of the year, and often incorporates two games per week, it is hard to maintain physical fitness at its peak throughout. Yet the maintenance of physical fitness during a season is a key target for every team (Koutedakis, 1995). The remaining three months are equally split between the off-season (or recess period) at the end of the competitive year and the pre-season training period before the start of the next annual league competition. Traditionally the off-season was deemed to be a dead period with players tending to return for pre-season training overweight (White *et al.*, 1988). The lack of training in the recess period would also lead to a deterioration in aerobic fitness and a fall in musculoskeletal strength. This decline compromises physical training pre-season and can be partly offset by maintaining fitness on an individual basis, and attending to dietary intake and weight control. Body composition is an important factor in preparing for peak performance in soccer. Excessive adipose tissue acts as a dead weight in activities where the body mass must be repeatedly lifted against gravity during locomotion and jumping (Reilly, 1996), whilst lean mass contributes to the production of power during high-intensity activity.

Recent technological advances have increased the precision and accuracy of body composition measurement by means of dual-energy x-ray absorptiometry (DEXA). The development of DEXA as a body composition tool has promoted a renewed interest in the field of anthropometry due to its unique ability to subdivide the body and its segments into separate components of bone mineral mass, fat mass and fat-free mass. In addition it overcomes the population-specific nature of equations for predicting body fat from anthropometric measures, and the assumptions of constant fat-free tissue density associated with hydrodensitometry.

There is a lack of recent information available on changes in body composition of professional soccer players during the whole season and just as importantly during the off-season, using accurate and high quality scanning images that are now available. We have already investigated at our laboratory pre-season changes using the same group and protocol (Egan *et al.*, 2006). The purpose of the present study was to establish the changes that occur in body composition during an off-season lay-off period in professional soccer players using DEXA.

2 MATERIALS AND METHODS

2.1 Participants

Participants were male professional English Premier League association football (soccer) players (n = 14; goal-keepers: 3; backs: 6; forwards: 3; midfield: 2), who underwent body composition assessment as part of an on-going routine assessment. The mean ± SD characteristics were: Age: 27.93 ± 4.84 years; Mass: 85.58 ± 6.60 kg; Height: 1.86 ± 0.63 m. Written informed consent was received from all players prior to testing.

2.2 Measurement protocol

Duplicate DEXA measurements (Figure 9.1) were made on the same players, during the late 2004 competitive season (March 2004) and repeated at the start of pre-season training (July 2004; week 1 of 6). No data were collected at the end of the previous season, as the club's support staff did not deem it an important time point for measurement of body composition due to the view that seasonal competitive performance would not subsequently be affected. No specific information was provided by the players regarding their off-season activity patterns, due to concerns of it being reported back to the coaching staff. All players were allowed to self-direct training in their recess phase; the coaching staff placed considerable emphasis on the maintenance of fitness during this break and provided guidelines for training but had no means of monitoring the activity intervention.

Matches were distributed relatively evenly across the competitive season, with games each week-end. When two matches were played within the same week, the training load was reduced to take this extra fixture into account.

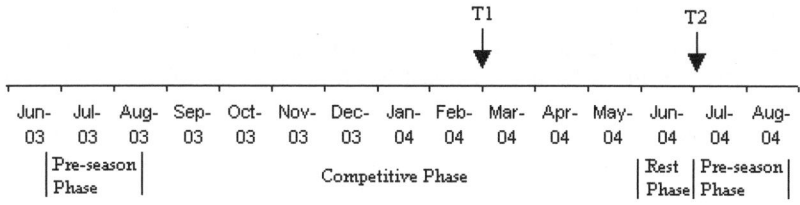

Figure 9.1 Time graph for measurement collection.

2.3 Body composition measurement

Subjects arrived at the laboratory in the morning, after fasting for three hours. Subjects' age, height (m), and body mass (kg) on beam balance scales were measured. Whole-body mass, fat mass, lean mass, percent body fat (%BF), bone mineral content (BMC) and bone mineral density (BMD) were then measured according to standard operating procedures using a fan beam dual-energy X-ray absorptiometry scanner (DEXA; Hologic QDR series, Delphi A, Bedford, Massachusetts). Scans were analysed using system Hologic QDR Software for Windows version 11.2 (© 1986–2001 Hologic Inc.) according to standard analysis protocols. Regional fat mass and fat-free soft tissue measurements were derived from total-body scans, using standard analysis methods. Subsequently, body composition values for leg (left leg + right leg), trunk (left ribs + right ribs + lumbar + thoracic spine + pelvis), and arm (left arm + right arm) regions were calculated. All scans were analysed by the same investigator.

Participants wore lightweight shorts without zips, buttons, or any other metal and removed all jewellery for the scanning. The procedures received approval from the University's Human Ethics Committee.

The coefficients of variation (CV) (Equation 9.1) of repeated total whole-body DEXA measurements in our laboratory ranged between 0.71% and 1.25% for total mass, fat mass, lean mass, percent body fat, bone mineral content and bone mineral density. Segmental variation was less than 4% for fat mass and less than 3% for lean mass. These values, together with the technical error of measurements (TEM; equation 2), are tabulated in Table 9.1.

$$\text{CV} = \frac{\left(\dfrac{SD_{diff}}{\sqrt{2}}\right)}{\overline{X}} \, x100 \qquad (9.1)$$

Where SD_{diff} is the standard deviation of the total differences between the measures; and $\overline{X}$ is the mean of all values.

$$\text{TEM} = \sqrt{\frac{\Sigma d^2}{2n}} \qquad (9.2)$$

Where d is the difference between each pair of measures; and n is the sample number.

2.4 Analysis

Results are expressed as mean ± SD where relevant. Changes in body composition (total mass, lean mass, fat mass, percent body fat and bone mineral content) and changes in bone density variables between the visits during the final stages of the season and the beginning of pre-season were made using Student's paired t-tests for matched samples. Statistical analysis was carried out using SPSS version 12.0.1. Results were considered to be significant when $P < 0.05$.

Table 9.1 Reliability measures for DEXA body composition assessment previously determined in our laboratory (n = 30).

Variable		CV (%)	TEM
Fat mass (kg)	Whole-body	1.16	0.20
	Left arm	2.71	0.03
	Right arm	3.99	0.04
	Trunk	2.84	0.19
	Left leg	2.23	0.08
	Right leg	2.45	0.09
	Total arms	2.31	0.04
	Total legs	1.72	0.13
Lean mass (kg)	Whole-body	0.71	0.30
	Left arm	2.74	0.05
	Right arm	3.00	0.06
	Trunk	1.14	0.25
	Left leg	2.05	0.14
	Right leg	1.71	0.12
	Total arms	1.98	0.08
	Total legs	1.62	0.22
Fat (%)	Whole-body	1.25	0.34
BMD (g·cm^2)	Whole-body	0.74	0.01
Mass (kg)	Whole-body	0.39	0.24

3 RESULTS

3.1 Body composition

Group mean ± SD body composition values for all the subjects tested at each time point are reported in Table 9.2. Following an off-season break from any official training of two months, there were no significant changes in any body composition component. The percent body fat as measured by DEXA did not increase over the off-season period (13.2 ± 2.0%–13.4 ± 1.3%; $P > 0.05$).

When analysed according to body segment (Figure 9.2), the changes we observed between the first and second visits were that there was a significant ($P < 0.05$) decrease in lean mass of the trunk and a significant ($P < 0.05$) increase in the lean mass of the arms, along with a non-significant increase in the lean mass of the legs. No overall change was noted for lean mass, due to the opposing changes in lean mass of the arms and trunk. Fat mass did not change significantly over the off-season period at any of the segments (Figure 9.2 and Table 9.3).

Table 9.2 Body composition variables (Mean ± SD).

Variables	Late Season	Beginning of Pre-season
Total Mass (kg)	87.15 ± 5.92	86.91 ± 5.68
Fat Mass (kg)	11.57 ± 2.19	11.65 ± 1.45
Lean Mass (kg)	71.82 ± 4.48	71.53 ± 4.59
Body Fat (%)	13.2 ± 2.0	13.4 ± 1.3
BMD (g.cm^2)	1.462 ± 0.11	1.453 ± 0.10

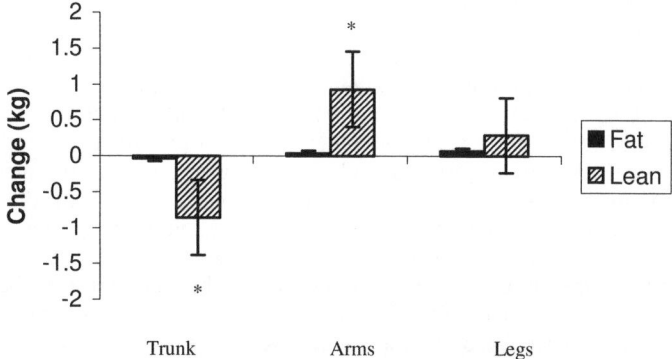

Figure 9.2 Changes (kg) in fat and lean mass for trunk, arm and leg regions between the two test sessions with standard error bars. Statistical significance (P < 0.05) changes are indicated by *.

Table 9.3 Regional values for fat mass and lean mass for the two assessment times (Mean ± SD).

	Fat Mass (kg)		Lean Mass (kg)	
	Late season	Beginning pre-season	Late season	Beginning pre-season
Trunk	4.33 ± 1.01	4.29 ± 0.73	35.24 ± 2.23	34.32 ± 2.18
Arms	1.37 ± 0.25	1.41 ± 0.20	8.41 ± 0.99	9.34 ± 1.03
Legs	4.67 ± 1.05	4.75 ± 0.91	25.42 ± 2.22	25.13 ± 2.27

3.2 Bone density

The measurement of whole-body bone mineral content indicated that players did not have any significant ($P > 0.05$) alterations in the bone mineral density (1.462 ± 0.11 g.cm^2 – 1.453 ± 0.10 g.cm^2), during the off-season period.

4 DISCUSSION

In a review of the anthropometry of elite soccer players Stolen *et al.* (2005) concluded that there have been no changes in height and mass of soccer players in recent decades. This stability contradicts data for Rugby Union international players, Norton and Olds (2001) referring to this trend as the 'expanding universe'. The soccer players in this study were over 5 cm taller and almost 10 kg heavier than those reported by Casajús (2001) and Ostojic (2003). Whether the data in the current sample reflect this change in anthropometric features in elite soccer players is too early to establish.

In sports where body mass has to be repeatedly moved against gravity any excess body fat acts as an undesirable load. The maintenance of an optimal body composition is necessary to retain an advantage over the opposition, throughout both a single match and throughout the season (Rico-Sanz, 1998). There are many methods and techniques used to assess body composition; use of callipers to measure skinfold thickness is among the most frequently adopted methods in assessment of soccer players, although there is not a well validated, soccer-specific prediction equation for professional male soccer players. The advancement of dual-energy X-ray absorptiometry, and a growing understanding of the need for physiological tests in laboratory conditions, have led to a greater number of professional clubs utilising this method. This method removes the errors associated with skinfold prediction equations and reduces the bias and error of the assessor, while providing additional information regarding fat-free soft tissue mass and bone mineral density, as well as providing a region/segmental breakdown.

Traditionally during the off-season period, most soccer players increased their fat content (White *et al.*, 1988) and body mass, presumably because of reduced aerobic activity and nutritional and behavioural changes. With players, even at high levels, tending to have body fat percent values higher than optimal (Bangsbo and Mizuno, 1988) it has became common practice to advise players to keep their activity profile relatively high during the off-season in an aim to stay fit and to prevent increased adiposity and muscle atrophy.

The repeated measurement of the body composition in our current sample of Premier League soccer players, between the end of the competitive season and again at the beginning of the pre-season training periods failed to demonstrate any significant changes in fat or fat-free components. In previous decades substantial changes in body fat had been observed during this off-season period (Burke *et al.*, 1986; White *et al.*, 1988). White *et al.* (1988), using skinfold measurements, reported mean body fat percent values of 19.3% at the start of pre-season training. Comparison with our current findings suggests that almost 20 years on, players are returning to the pre-season training programme in a well maintained physical condition, having adopted a self-directed fitness maintenance programme throughout the off-season, as a result of pressure from the coaching staff to maintain fitness during their break. However, it was not possible to quantify the players' physical activity levels during this break, in the current study. This practice allows for improvement rather than restorative pre-season training, so players can begin the season in a superior physical condition. However, some recent studies still disclosed that the body fat content of elite professional soccer players significantly dropped during the conditioning and competitive period but significantly increased during the off-season (Ostojic, 2003). It seems therefore that the philosophy of physical training in the off-season has not been adopted in all countries.

Despite the stability in whole-body composition values, lean mass changes were evident in all anatomical regions, although variable. Gains in lean mass of the arms and legs and a loss in lean mass of the trunk were observed. This patterning of lean mass over the season would require further long-term monitoring for replication. There was no significant change in fat mass in any of the body segments. While the ability of DEXA to divide the body into segmental regions has the potential to provide a useful insight into regional fat and lean tissue distribution and changes over time, it should be noted that reliability of sub-regional measurements is less than for whole-body scans due to the introduction of human error in the positioning of divisional lines between the body segments during analysis and the smaller masses involved.

5 CONCLUSIONS

The magnitudes of the body composition changes in professional soccer players during the off-season were largely non-significant. Within the confines of this study, the values indicate that the soccer players monitored did not show significant gains in fat mass, lean mass or lose bone mineral content during their off-season break, the players' total mass remained unchanged after the six week vacation. This maintenance of body composition allows players to return to pre-season training better prepared for the formal conditioning work it entails.

The regional body compositional results should be of interest to soccer coaches, because they may help direct where the strength and conditioning work needs to be targeted during the pre-season training period. Periodic measurement of body fat percentage and fat-free soft tissue allows the coaches and support staff to correct the training programme and diet as deemed necessary.

6 REFERENCES

Bangsbo, J. and Mizuno, M., 1988, Morphological and metabolic alterations in soccer players with detraining and retraining and their relation to performance. In *Science and Football*, edited by Reilly, T., Lees, A., Davids, K. and Murphy, W.J., London E. & F.N. Spon, pp. 114–124.

Burke, L.M., Gollan, R.A. and Read, R.S., 1986, Seasonal changes in body composition in Australian Rules footballers, *British Journal of Sports Medicine*, **20**, 69–71.

Casajús, J.A., 2001, Seasonal variation in fitness variables in professional soccer players. *Journal of Sports Medicine and Physical Fitness*, **41**, 463–467.

Egan, E., Reilly, T., Chantler, P. and Lawlor, J., 2006, Body composition before and after six weeks pre-season training in professional football players. In *Kinanthropometry IX; Proceedings of the 9th International Conference of the International Society for the Advancement of Kinanthropometry (ISAK)*, edited by Marfell-Jones, M. and Stewart, A. (London: Routledge), pp. 123–130.

Koutedakis, Y., 1995, Season variation in fitness parameters in competitive athletes. *Sports Medicine*, **19**, 373–392.

Norton, K. and Olds, T., 2001, Morphological evolution of athletes over the 20th Century: Causes and consequences. *Sports Medicine*, **31**, 763–783.

Ostojic, S.M., 2003, Seasonal alterations in body composition and sprint performance of elite soccer player. *Journal of Exercise Physiology*, **6**, 11–14.

Reilly, T., 1996, Fitness assessment. In *Science and Soccer* edited by Reilly, T., (London: E. & F. Spon), pp. 25–50.

Rico-Sanz, J., 1998, Body composition and nutritional assessment in soccer. *International Journal of Sport Nutrition*, **8**, 113–123.

Stolen, T., Chamari, K., Castagna, C. and Wisloff, U., 2005, Physiology of soccer; an update. *Sports Medicine*, **35**, 501–536.

White, J.E., Emery, T.M., Kane, J.E., Groves, R. and Risman, A.B., 1988, Pre-season fitness profiles of professional soccer players. In: Science and Football, edited by Reilly, T., Lees, A., Davids, K. and Murphy, W.J., London E. & F.N. Spon, pp. 164–171.

CHAPTER TEN

The relationship between strength, power, flexibility, anthropometry and technique and 2000 m and 5000 m rowing ergometer performance

P. Graham-Smith, K. Burgess and A. Ridler

Centre for Rehabilitation and Human Performance Research, University of Salford, Greater Manchester, United Kingdom.

1 INTRODUCTION

The sport of rowing requires participants to move a boat's shell over a given distance (most major competitions use 2000 m) in the quickest time possible (Smith and Loschner, 2002). Rowers are regularly subjected to performance testing to assess their current level of fitness and to determine their ranking within the squad with whom they compete for boat selection. This testing is commonly carried out on rowing ergometers as water testing is often more susceptible to the influences of other variables affecting the outcomes (such as weather, water conditions, equipment used, and crew combinations). Race situations are replicated by performing similar distances to that of competitions (most commonly 2000 and 5000 m). Studies into the accuracy of these rowing ergometers have concluded that the Concept II ergometer is a valid and reliable measure of on-water performance (Macfarlane et al., 1997; Schabort et al., 1999; Soper and Hume, 2004).

However, when such tests are used for talent identification purposes they are time consuming and novice rowers often find efforts over 2000 m and 5000 m physiologically demanding. Additionally, their actual potential may not be accurately assessed due to the requirement of good technique. Instead governing bodies have adopted a number of shorter field tests to assess the potential of new recruits to schemes such as the Amateur Rowing Association's (ARA) 'World Class Start' programme. Here novice rowers are required to undergo tests for strength, power, flexibility and also to have simple anthropometric measurements taken. Despite the physiological demands of the sport being well documented, i.e. observations that rowers require high levels of both aerobic and anaerobic power (Shepherd, 1998), and both good strength and endurance (McGregor et al., 2002), such a battery of tests can be regarded as more biomechanical than physiological because they do not assess aerobic power and muscular endurance.

Although these tests are common (and indeed new devices such as the Concept II® DYNO have been developed to measure rowing-specific strength), there are few investigations which test the validity of such measures against actual rowing performance. Surprisingly few studies have examined aspects of rowing technique alongside physical and anthropometric measurements. Associations between the field test results and rowing performance are often made in isolation, i.e. multiple single correlations, and analyses have not progressed into examining the combined effect of variables on reducing the amount of unknown variance in rowing times.

The aim of this paper is, therefore, to examine rowing performance from a biomechanical perspective and to examine the relationship between physical, anthropometric and technical factors on 2000 m and 5000 m rowing performance. The next section outlines an 'a priori model' focussing on each of these factors in turn, highlights relevant variables and provides a rationale for their inclusion in the study.

1.1 *A-priori* model

1.1.1 *Physical factors*

The motion of rowing is one which recruits many major muscle groups. Rowing requires the transfer of power developed at the catch (where the blade enters water) by the foot on the footplate, effectively through the body, until the hands perform the finish (point where the blade leaves water) of the stroke (Baudouin and Hawkins, 2002). In light of these simple requirements it is obvious that the ability to generate force is a very important variable in the speed an oarsman can produce in the boat.

Although rowing is a weight supported sport Shephard (1998) observed that the resistance to forward motion is proportional to the weight of the boat and its crew members (to the power of approximately two thirds). With this in mind it is possible to argue that the strength of a rower is important, as the more force the rower can generate through the oar, the greater the boat's propulsion. Baudouin and Hawkins (2002) reported that the forces applied to the oar are primarily a result of the force developed at the foot stretcher through the leg musculature. They also acknowledged that the maximum effect of the leg force generation will only be attained if rowers have good musculoskeletal strength across all joints allowing for sequential phasing and efficient force transfer to the oar.

Kramer *et al.* (1994) found that leg strength was a better discriminator of 2500 m rowing performance than other measurements of strength conducted on the arm, back and chest. The importance and significance of leg strength in rowers was also reported by Parkin *et al.* (2001) who observed that rowers have significantly stronger quadriceps during knee extension than non-rowers.

Although maximum strength (force) is undoubtedly important, boat speed is also related to the acceleration (and therefore velocity) in which the oar can be pulled through the water. Power is therefore an important physical attribute that an oarsman must possess. Peak power has been shown to be strong predictor of 2000 m rowing performance ($R^2 = 0.86$, Ingham *et al.*, 2002); and peak power per kilogram of body mass was found to be an excellent discriminator between novice, good, and elite oarsmen (Smith and Spinks, 1995). Typical force and power outputs for rowers were provided by Hartmann *et al.* (1993) who observed that world class male oarsmen produced mean peak force and power of 975 N and 3230 W respectively in the fifth stroke of a ten-stroke maximum test.

However, with an increase in strength, increases in muscle mass are expected. As rowing requires the oarsmen to propel the shell of a boat as quickly as possible, any increases in weight will cause the boat to sit lower in the water, thus increasing resistance acting on the hull. Therefore, any increase in resistance is only beneficial if propulsive force gained is greater than the increase in resistance (De Rose *et al.*, 1989) with the balance of power to body weight affecting the speed and smoothness experienced by the boat (Smith and Spinks, 1995).

Flexibility, particularly of the musculature surrounding the lower back and legs, is another physical factor that has the potential to improve or reduce a rower's range of motion and therefore affect technical factors such as stroke length. Therefore, to represent the physical aspects of rowers we chose to measure aspects of strength (maximum and average forces and work done in bench press, bench pull and leg press movements), peak power and flexibility.

1.1.2 Anthropometric factors

In a study by Bourgois *et al.* (2001) it was reported that World Junior Rowing Championship finalists were taller, heavier, with greater limb length, breadth, and girth than semi-finalists. These characteristics, particularly limb length, allow the oarsman greater leverage. De Rose *et al.* (1989) also indicated that larger rowers posses larger muscle cross sectional area and thus a greater metabolic capacity.

Studies by Russell *et al.* (1998) and Cosgrove *et al.* (1999) investigated how different variables (physiological, anthropometric, and strength) predict rowing performance determined by a 2000 m ergometer test. Both reported that anthropometric measurements offered the most significant correlations with 2000 m performance, particularly height ($R^2 = 0.61$, Russell *et al.*, 1998), and body mass and lean muscle mass ($R^2 = 0.48$ and 0.71 respectively, Cosgrove *et al.*, 1999).

Yoshiga and Higuchi (2003a) also reported that height was significantly correlated with ergometer performance ($R^2 = 0.23$) as was body mass, lean body mass and bilateral leg extension power ($R^2 = 0.53$, 0.58 and 0.38 respectively). In multiple regression analysis a combination of lean body mass and leg extension power raised the coefficient of determination to an R^2 value of 0.63.

To represent the anthropometry of rowers we chose to measure height and body mass.

1.1.3 Technical factors

The studies above have not taken into account technical factors, many of which have been shown to discriminate between elite and sub-elite rowers (Smith and Loschner, 2002; Baudouin and Hawkins, 2002; Soper and Hume, 2004). Several very simple temporal and displacement measurements have been used to adequately represent differences in technique. These include stroke length, drive time, and recovery time, with the latter two being used to calculate the recovery to drive ratio. Ultimately boat velocity is the product of stroke length and stroke velocity (O'Neill, 2004), but the timing of the drive and recovery also play an important role in maintaining the momentum of the boat itself.

Smith and Loschner (2002) and Soper and Hume (2004) both stressed the importance of a smooth recovery and the ratio of recovery time relative to drive time. At the end of the drive the oars are lifted from the water and the impulse generated by the oars is transferred into the speed of the boat. The boat moves at its fastest speed during the recovery phase so it essential that the rower moves smoothly during the recovery, avoiding sudden movements as this would cause the boat to jerk and roll, increasing the resistance acting on the hull, and reduce its speed.

A recovery to drive ratio of 2:1 is often referred to as desirable (Redgrave, 1995), but in reality it is often within the range of 0.9 to 1.7 during analysis of rowing performance (Soper and Hume, 2004). A ratio where recovery time is equal to or less than drive time is considered detrimental to the average boat speed. Although a fast recovery (acceleration forwards) of the rower to the catch position creates an acceleration of the boat (due to the conservation of momentum) a series of large accelerations and decelerations has a detrimental affect on the average boat speed. It is considered more beneficial and efficient to return slower to the catch and prolong the glide of the boat when it is travelling its quickest at the end of the stroke (O'Neill, 2004). Considering a good crew accelerates the boat to travel 1.5 times further during the recovery compared to the drive (O'Neill, 2004), it is essential to achieve a recovery where the boat is allowed to glide off the finish. This will allow the boat to continuously move forward, rather than be moved by a series of jerking movements.

Investigations into 2000 m rowing performance have also highlighted the importance of stroke length to rowing performance. Ingham *et al.* (2002) reported stroke length to be highly significant to 2000 m rowing performance accounting for 29% of the variance in rowing performance ($R^2 = 0.29$), whilst a report by Secher (1983) concluded that taller oarsman have an advantage as they have greater height to secure a longer stroke length.

We represented the technical aspects of rowing with four measurements: stroke length, drive time, recovery time and the recovery: drive time ratio.

2 METHODS

2.1 Subjects

Eighteen male oarsmen were selected from a convenience sample of a representative population of rowers taken from University and local rowing clubs. All participants had at least two years' rowing experience and were familiar with performing maximal rowing efforts over 2000 m and 5000 m on a Concept II® rowing ergometer and all were healthy and free from injury at the time of testing.

Each rower underwent a battery of fitness tests devised to assess several physical, physiological, anthropometric and technical attributes in order to determine their association with rowing performance over 2000 m and 5000 m. These tests were selected on the basis of them being routinely used for athlete screening and they were as specific to the demands and movements patterns of rowing as possible.

For ease of explanation these tests will be described in relation to the equipment and attributes they were designed to measure.

2.2 Concept II® (Model D) Rowing Ergometer tests

All tests that required the Concept II rowing ergometer (i.e. endurance and peak power) were set at the same resistance for all athletes, equating to a drag factor of 140. Foot stretcher positions were self selected by the athlete according to their individual preferences.

2.2.1 Endurance

Endurance and overall rowing performance were determined via two maximal effort time trials over 2000 m and 5000 m distances. Athletes were requested to remain within a stroke rate range of 28 to 35 strokes per minute to control for any extraneous effects on factors such as power, torque and stroke length. These tests were conducted one week apart and no maximal intensity training was performed on the day preceding these tests. The times taken to cover these distances were our performance (dependent) variables in the statistical analyses and these were expressed in seconds.

2.2.2 Power

Peak power was established via a five stroke maximum test whereby the athlete performed two build up strokes before pulling five maximum effort strokes on the ergometer (Ingham *et al.*, 2002).

2.3 Concept II DYNO® Strength tests

Rowing specific strength was assessed using a Concept II DYNO® to measure maximal bench press, leg press and bench pull strength. The DYNO® was set at fu" resistance, and following three warm up repetitions the athletes performed thre naximal efforts, taking 5 seconds to slowly return back into the start position (this s the recommended protocol to maintain inertial loads and avoid 'pumping'). The maximum and average scores (in kg) and work done (in Nm) were recorded for each of the three movements.

2.4 Anthropometric tests

Basic anthropometric measures of height and body mass were carried out before all other testing took place. Height was measured to the nearest 0.1 cm from the top of the cranium to the floor with the subject standing unshod and upright against a wall. Body mass was measured using SECA® scales and recorded to the nearest 0.1 kg.

2.5 Flexibility

The sit and reach test was used to measure the athlete's flexibility in their lower back and hamstrings. With the athlete sitting on the floor with unshod feet against the sit and reach box, they were instructed to keep both legs straight such that the back of their knees always remained in contact with the ground. The tips of the first finger were touching as the athlete reached forward as far as possible onto the top of the sit and reach box. Athletes were given three attempts and the highest score (to the nearest cm) was taken for further analysis.

2.6 Technique analysis

At the 500 m mark in the performance of the 2000 m time trial ten strokes of each athlete were filmed in the sagittal plane. The Sony® (DCR-TVR950E) video camera was positioned 6 m away from and in line with the centre of the Concept II rowing ergometer, giving a horizontal field of view of 3 m. The length and height of the ergometer provided horizontal and vertical reference scales for calibration purposes. Quintic® software (version 9.03) was used to capture and analyse the video footage at 50 frames per second. Two critical instants in each stroke were identified; the 'catch' and the 'end of drive'. The catch was defined as the first frame where the handle showed a definite move away from the flywheel (see Figure 10.1); the 'end of pull' was defined as the instant where the handle had momentarily come to rest reaching its furthermost position away from the flywheel (see Figure 10.2). Using these reference points it was possible to determine the following four technical variables as descriptors of technique:

(i)   stroke length – defined as the distance travelled by the handle at 'catch' to the furthermost position at the end of the pull (see Figure 10.2)

(ii)   stroke (drive) time – defined as the time interval between the instants of 'catch' and 'end of pull'

(iii)   recovery time – defined as the time interval between 'end of pull' and the 'catch' of the following stroke, and finally

(iv)   recovery to drive time ratio was calculated (by dividing recovery time by drive time).

For each athlete ten consecutive strokes were analysed and the average of these four variables was taken for subsequent statistical analysis.

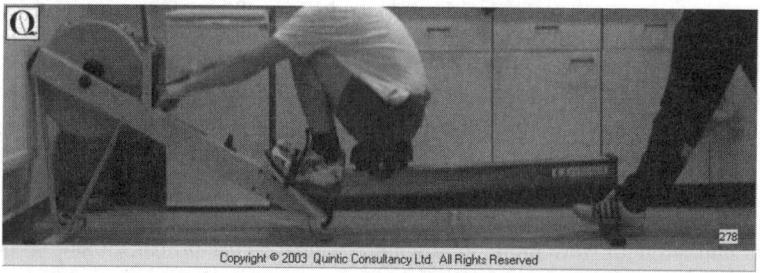

Figure 10.1 The 'catch' position denoting the instant that the stroke (drive) is initiated.

Figure 10.2 The furthermost position of the handle denoting the 'end of drive'. The line connecting 'catch' and 'end of drive' positions depicts the stroke length.

2.7 Statistical analysis

Data were analysed in Minitab statistical software (version 14) to generate basic descriptive statistics and to examine relationships between anthropometric, physical and technical variables with rowing performance over 2000 m and 5000 m (in accordance with our *a-priori* model outlined in the introduction). Relationships between variables were tested using the Pearson Product Moment Correlation

Coefficient and the coefficient of determination (R^2). 'Best subsets' multiple regression analysis was then performed to investigate whether combinations of anthropometric, physical and technical variables led to better predictions of rowing performance over 2000 m and 5000 m, thereby attempting to reduce the amount of unknown variance. A maximum of three predictor variables were permitted in accordance with the recommendations of Vincent (1995) relating to a minimum of five subjects per variable.

3 RESULTS

Descriptive statistics for all the variables measured are presented in Table 10.1. All variables exhibited a normal distribution. Rowing performances over 2000 m and 5000 m equate to times of 6 min 35 s $\pm$ 4 s and 17 min 26 s $\pm$ 9 s respectively.

The correlation analysis between the performance variables (2000 m and 5000 m times) and the predictor variables (anthropometric, physical and technical variables) revealed that over both distances the athlete's height and peak power produced the strongest correlations (all $p < 0.000$, see Table 10.2). When tested in isolation the athlete's height explained 60% and 50% of the variation in 2000 m and 5000 m rowing times, whilst peak power explained 76% and 75% respectively.

Of the strength measurements it was not surprising that the leg press and bench pull strength were more significantly related to performance than the bench press strength as these are more specific to movement patterns in rowing. Average strength measures over the three repetitions were slightly better predictors of performance than the maximum scores, but interestingly it was the work done in all three movements that was the most significant, demonstrating coefficients of determination between 36% and 47% (all $p < 0.01$).

In terms of technique-related variables, stroke length was found to be the only significant variable explaining 45% and 51% of the variances in 2000 m and 5000 m rowing times ($p = 0.02$ and 0.001 respectively).

Flexibility was not found to be a significant factor in either 2000 m or 5000 m rowing times.

Multiple regression analysis found small improvements in raising the coefficient of determination (R^2 adjusted) from 75.9% (peak power alone versus 2000 m time) to 76.6% (combined with recovery:drive ratio) and 76.3% (combined with stroke length). However, both these additional variables fell short of being significant contributors, exhibiting p-values of 0.132 and 0.15 respectively.

For the 5000 m analysis, stroke length was found to be a significant contributor ($p = 0.05$) when combined with peak power, raising the coefficient of determination from 75.3% (peak power alone) to 78.4%, thereby decreasing the amount of unknown variance to 21.6% ($F = 31.8$, $p = 0.001$). In real terms this additional variable helped to improve the predictability of the regression equation from a standard error of 20.5 seconds to 18.6 seconds. The multiple regression equation derived was:

$$5000 \text{ m time} = 1611 - 0.459 \text{ Peak Power} - 114 \text{ Stroke Length}$$

Table 10.1 Descriptive statistics of the variables tested.

Variables		Mean	SD	Min	Max
Performance Measures					
2000 m time	(s)	395.1	16.1	367	417
5000 m time	(s)	1046.9	40.1	971	1113
Anthropometry					
Height	(cm)	186.0	7.5	173	201
Body Mass	(kg)	85.3	9.6	65	102
Flexibility					
Sit & Reach	(cm)	21.8	9.1	8	49
Strength					
Bench Press Maximum	(kg)	81.2	10.4	65	104
Bench Press Average	(kg)	79.2	10.0	64	103
Bench Press Work Done	(Nm)	406.1	56.8	304	526
Leg Press Maximum	(kg)	162.2	17.6	134	197
Leg Press Average	(kg)	158.4	17.6	128	193
Leg Press Work Done	(Nm)	824.5	130.4	658	1074
Bench Pull Maximum	(kg)	85.6	12.1	59	108
Bench Pull Average	(kg)	82.7	11.4	58	105
Bench Pull Work Done	(Nm)	547.0	137.0	315	845
Power					
Peak Power	(W)	792.7	59.9	721	897
Technique					
Stroke Length (m)	(m)	1.76	0.11	1.62	2.04
Drive time (s)	(s)	0.88	0.05	0.78	0.98
Recovery time (s)	(s)	1.21	0.25	0.98	1.92
Recovery:Drive ratio		1.37	0.24	1.09	2.00

Table 10.2 Correlations and coefficients of determination between 2000 m and 5000 m ergometer performance with predictor variables.

Variables		2000 m			5000 m		
		r	p	R^2	r	p	R^2
Performance							
2000 m time	(s)	–	–	–			
5000 m time	(s)	0.96	0.000	0.92	–	–	–
Anthropometry							
Height	(cm)	–0.78	0.000	0.60	–0.74	0.000	0.55
Body Mass	(kg)	–0.54	0.021	0.29	–0.55	0.018	0.30
Flexibility							
Sit & Reach	(cm)	–0.34	0.173	0.11	–0.39	0.106	0.16
Strength							
Bench Press Maximum	(kg)	–0.41	0.094	0.17	–0.36	0.139	0.13
Bench Press Average	(kg)	–0.42	0.079	0.18	–0.39	0.115	0.15
Bench Press Work Done	(Nm)	–0.69	0.002	0.47	–0.66	0.003	0.44
Leg Press Maximum	(kg)	–0.46	0.058	0.21	–0.36	0.138	0.13
Leg Press Average	(kg)	–0.48	0.045	0.23	–0.38	0.122	0.14
Leg Press Work Done	(Nm)	–0.65	0.003	0.43	–0.61	0.008	0.37
Bench Pull Maximum	(kg)	–0.59	0.009	0.35	–0.53	0.024	0.28
Bench Pull Average	(kg)	–0.64	0.004	0.41	–0.58	0.011	0.34
Bench Pull Work Done	(Nm)	–0.65	0.003	0.43	–0.67	0.002	0.45
Power							
Peak Power	(W)	–0.87	0.000	0.76	–0.87	0.000	0.75
Technique							
Stroke Length (m)	(m)	–0.67	0.020	0.45	–0.72	0.001	0.51
Drive time (s)	(s)	0.10	0.690	0.01	0.03	0.904	0.00
Recovery time (s)	(s)	0.32	0.193	0.10	0.19	0.448	0.04
Recovery:Drive ratio		0.34	0.172	0.11	0.22	0.392	0.05

The fact that peak power was by far the most significant variable in predicting 2000 m and 5000 m performance led us to investigate whether the other physical factors, anthropometric measures and technical aspects were related to power production. The results in Table 10.3 show significant relationships with all anthropometric and physical variables, but out of the four technical variables only stroke length was significant.

Table 10.3 Relationships between peak power and anthropometric, physical and technical measures.

Variables		Peak Power		
		r	p	R^2
Anthropometry				
Height	(cm)	0.82	0.000	0.67
Body Mass	(kg)	0.69	0.001	0.48
Flexibility				
Sit and Reach	(cm)	0.51	0.030	0.26
Strength				
Bench Press Maximum	(kg)	0.54	0.021	0.29
Bench Press Average	(kg)	0.56	0.015	0.32
Bench Press Work Done	(Nm)	0.77	0.000	0.59
Leg Press Maximum	(kg)	0.51	0.032	0.26
Leg Press Average	(kg)	0.51	0.032	0.26
Leg Press Work Done	(Nm)	0.79	0.000	0.62
Bench Pull Maximum	(kg)	0.63	0.005	0.40
Bench Pull Average	(kg)	0.69	0.002	0.47
Bench Pull Work Done	(Nm)	0.77	0.000	0.59
Technique				
Stroke Length (m)	(m)	0.61	0.007	0.37
Drive time (s)	(s)	−0.22	0.385	0.05
Recovery time (s)	(s)	−0.21	0.396	0.05
Recovery:Drive ratio		−0.18	0.486	0.03

4 DISCUSSION

The reliability and validity of rowing ergometers have been evaluated many times in the literature (particularly the Concept II[®]) and they have been shown to be both reliable (Macfarlane *et al.*, 1997) and an accurate and valid measure of on-water performance (Schabort *et al.*, 1999; Soper and Hume, 2004). Due to the fact that on-water rowing analysis is very expensive and impractical for this investigation, a Concept II[®] model D ergometer was used to test the validity of specific field tests that are commonly used for crew selection and for talent identification purposes and predicting the potential of novice rowers.

The results of this study support previous research findings demonstrating that rowing performance is highly associated with peak power (Hartmann *et al.*, 1993; Smith and Spinks, 1995; Ingham *et al.*, 2002; Yoshiga and Higuchi, 2003a); body height (Russell *et al.*, 1998; Cosgrove *et al.*, 1999; Yoshiga and Higuchi, 2003a, b); leg strength (Kramer *et al.*, 1994) and stroke length (Ingham *et al.*, 2002). Of these physical, anthropometric and technical variables, peak power was found to be the most influential factor accounting for over 75% of the variance in 2000 m and 5000 m rowing times.

Of the anthropometric measures body height was a better indicator of performance than body mass, accounting for 60% and 29% of the variance in 2000 m times respectively. The importance of body height (and longer limb lengths) cannot be overstated as height was found to be highly related to peak power ($R^2 = 0.67$) and has obvious benefits to producing longer strokes Secher (1983). Stroke length was the only technical variable found to be related to rowing performance.

In terms of strength measures it was interesting to note that maximum strength in the bench pull exercise produced higher coefficients of determination with 2000 m and 5000 m rowing times than did maximal leg press strength which is contrary to the findings of Kramer *et al.* (1994) – (although leg strength was still found to be a highly significant attribute). Work done over the three repetitions in the bench press, bench pull and leg press were all much better indicators of rowing performance than maximal and average strength measurements. This again highlights the importance of body height (and associated longer limb lengths) allowing the rower to apply a force over a longer distance, thereby generating greater impulse to the oar. Average strength over the three repetitions produced slightly better relationships than maximal strength measurements, possibly implying a greater emphasis on muscular endurance rather than absolute strength.

Although it was not a great surprise to find that stroke length was associated with rowing performance, it was interesting to note that the temporal aspects of technique (drive time, recovery time and recovery:drive time ratio) all produced non-significant relationships with rowing times. Considering that the 18 rowers in this study were recruited from two local crews (and some rowed in both crews) it would not be unreasonable to argue that the timing of the drive and recovery should be harmonious. However, whilst the range in drive time results was only small (0.2 s), the recovery time produced a much wider range of results, 0.94 s.

This indicates that even within this sample temporal aspects are still quite individual, despite the rowers stroking within a fairly narrow range of between 28 to 35 strokes per minute. It also demonstrates that temporal aspects of strokes are not as critical on ergometers as they are on-water where they can have a much more profound affect on boat speed (O'Neill, 2004).

In terms of the combined effect of variables, the multiple regression analyses found that no additional variable could better the predictive ability of peak power for the 2000 m row. Over the longer distance of 5000 m stroke length became a significant contributor raising the coefficient of determination to 78.4%. Although peak power and stroke length show a strong relationship between them ($R^2 = 0.37$), it is not as significant a relationship as peak power exhibits with body height ($R^2 = 0.67$). This suggests that the taller rowers in this study may not be making the most of their mechanical advantage. One possible explanation is flexibility which is highlighted by a wide range in results and exhibited poor correlations with 2000 m and 5000 m rowing times.

The present study has adopted a biomechanical focus on predicting rowing ergometer performance and has attempted to evaluate the individual and combined relationships of physical, anthropometric and technical aspects on performance.

A number of attributes identified in the *a-priori* model were found to be highly related to rowing performance. Therefore, in terms of talent identification, it is not an unreasonable policy for governing bodies to search for tall, powerful and strong athletes. Implementation of these shorter tests will help avoid exposing novice rowers to the high physiological and technical demands of rowing 2000 m that they may not be appropriately conditioned for.

The 16 variables representing strength, power, flexibility, anthropometry and technique were selected on the basis of an *a-priori* model outlined previously. The effect size of the relationships (R^2) with 2000 m and 5000 m rowing times were deemed to be more meaningful than the actual level of significance as the coefficient of determination explains the amount of known (and unknown) variance between variables. In light of this the Bonferroni correction, used to avoid the risk of making type 1 errors, was not applied to the data as it does not affect the main findings of this paper. Ideally the multiple regression equation derived for 5000 m time (combining peak power and stroke length) needs to be validated from data collected on a similar subject group. However, the predictive capability of a combination of variables is secondary to the main purpose of this study: to identify how each of the anthropometric, physical and technical variables commonly used in talent identification schemes relate to performance on a rowing ergometer. Indeed, the inclusion of stroke length, although a significant contributor, only reduced the amount of unknown variance by 3%.

Whilst the predictive capability of these tests was very good, accounting for 78.4% of the variance in 5000 m rowing time, research also needs to take into account the additional effects of purely physiological measures to reduce the amount of unknown variance, e.g. lean body mass, aerobic capacity, lactate threshold/tolerance tests. Reichman *et al.* (2002) observed that 75.7% of the variance in 2000 m rowing time could be explained by power alone, but the inclusion of VO_2 max and fatigue in a 30 second Wingate test helped to reduce the unknown variance by 12.1% and 8.2% respectively. Lean body mass was found to be a better predictor of rowing performance than overall body mass (Russell *et al.*, 1998; Cosgrove *et al.*, 1999; Yoshiga and Higuchi, 2003b) as this gives a more clear expression of muscle force capability. Cosgrove *et al.* (1999) and Perkins and Pivarnik (2003) have both found significant relationships between 2000 m rowing performance and the workload corresponding to a blood lactate concentration of 4 mmol·l^{-1}, suggesting that rowing economy may also contribute to success.

5 CONCLUSION

The results indicate that power production over five strokes is by far the greatest predictor of rowing performance among our physical, anthropometric and technical measures. Body height was the second most influential factor which was also found to be associated with peak power. One technical consideration that was highlighted as a significant additional contributor to predicting 5000 m rowing performance was stroke length, again emphasising the importance of height and long limbs. The importance of height (and associated limb length) was further emphasised when the work done in all strength tests was found to elicit much greater relationships to performance than simple measures of maximal and average strength over three repetitions.

The physical, anthropometric and technical measures adopted in this study appear to be a valid and useful battery of tests that are well suited to identifying talented athletes without the need to undergo time consuming and invasive physiological tests in the first instance.

6 REFERENCES

Baudouin, A. and Hawkins, D., 2002, A biomechanical review of factors affecting rowing performance. *British Journal of Sports Medicine*, **36**, 396–402.

Bourgois, J., Claessens, A.L., Janssens, M., Van Renterghem, B., Loos, R., Thomis, M., Pkilippaerts, R., Lefevre, J. and Vrijens, J., 2001, Anthropometric characteristics of elite female junior rower. *Journal of Sports Science*, pp. 195–202.

De Rose, E.H., Crawford, S.M., Kerr, D.A., Ward, R. and Ross, W.D., 1989, Physique characteristics of Pan American games lightweight rowers. *International Journal of Sports Medicine*, **10**, 292–297.

Cosgrove, M.J., Wilson, J., Watt, D. and Grant, S.F., 1999, The relationship between selected physiological variables of rowers and rowing performance as determined by a 2000 m ergometer test. *Journal of Sports Sciences*, **17**, 845–852.

Hartmann, U., Mader, A., Wasser, K. and Klaver, I., 1993, Peak force, velocity and power during five and ten maximal ergometer strokes by world class female and male rowers. *International Journal of Sports Medicine*, **14**, 542–546.

Ingham, S.A., Whyte, G.P., Jones, K. and Nevill, A.M., 2002, Determinants of 2000 m rowing ergometer performance in elite rowers. *European Journal of Applied Physiology*, **88**(3), 243–246.

Kramer, J.F., Leger, A., Paterson, D.H. and Morrow, A., 1994, Rowing performance and selective descriptive, field, and laboratory variables. *Canadian Journal of Applied Physiology*, **19**(2), 174–184.

Macfarlane, D.J., Edmond, I.M. and Walmsley, A., 1997, Instrumentation of an ergometer to monitor reliability of rowing performance. *Journal of Sports Sciences*, **15**, 167–173.

McGregor, A.H., Anderton, L. and Gedroyc, W.M.W., 2002, The trunk muscles of elite oarsmen. *British Journal of Sports Medicine*, **36**, 214–217.

O'Neill, T., 2004, Basic rigging principles – an Oarsport production. Accessible through www.oarsports.com.

Perkins, C.D. and Pivarnik, J.M., 2003, Physiological profiles and performance predictors of a women's NCAA rowing team. *Journal of Strength and Conditioning Research*, **17**(1), 173–176.

Parkin, S., Nowicky, A.V., Rutherford, O.M. and McGregor, A., 2001, Do oarsmen have asymmetries in the strength of their back and leg muscles? *Journal of Sports Sciences*, **19**, 521–526.

Redgrave, S., 1995, *The Complete Book of Rowing, 2nd Edition*. London: Partridge press.

Reichman, S.E., Zoeller, R.F., Balasekaran, G., Goss, F.L. and Robertson, R.J., 2002, Prediction of 2000 m indoor rowing performance using a 30 s sprint and maximal oxygen uptake. *Journal of Sports Sciences*, **20**, 681–687.

Russell, A.P., Le Rossignol and Sparrow, W.A., 1998, Prediction of elite schoolboy 2000–m rowing ergometer performance from metabolic, anthropometric and strength variables. *Journal of Sports Sciences*, **16**, 749–754.

Secher, N.H., 1983, The Physiology of rowing. *Journal of Sports Sciences*, **1**, 25–53.

Schabort, E.J., Hawley, J.A., Hopkins, W.G. and Blum, H., 1999, High reliability of performance of well–trained rowers on a rowing ergometer. *Journal of Sports Sciences*, **17**(8), 627–632.

Shephard, R.J., 1998, Science and medicine of rowing: A review. *Journal of Sports Sciences*, **16**, 603–620.

Smith, R.M. and Loschner, C., 2002, Biomechanical feedback for rowing. *Journal of Sports Sciences*, **20**, 783–789.

Smith, R.M. and Spinks, W.L., 1995, Discriminant analysis of biomechanical differences between novice, good, and elite rowers. *Journal of Sports Sciences*, **13**, 377–385.

Soper, C. and Hume, P.A., 2004, Towards an ideal rowing technique for performance – the contribution from biomechanics. *Sports Medicine*, **34**(12), 825–848.

Vincent, W.J., 1995, *Statistics in Kinesiology*. Human Kinetics, Champaign, Illinois.

Yoshiga, C.C. and Higuchi, M., 2003a, Bilateral leg extension power and fat–free mass in young oarsmen. *Journal of Sports Sciences*, **21**, 905–909.

Yoshiga, C.C. and Higuchi, M., 2003b, Rowing performance of female and male rowers. *Scandinavian Journal of Medicine and Science in Sports*, **13**, 317–321.

CHAPTER ELEVEN

Anthropometric and physiological characteristics of elite female water polo players

K. Marrin and T.M. Bampouras

Edge Hill University, Sport and Exercise Research Group
St Helens Road, Ormskirk L39 4QP, UK

1 INTRODUCTION

Water polo is a non-contact sport consisting of 4 periods of 8 minutes actual playing time, separated by 2 minutes interval between periods and 5 minutes at half time. The intermittent nature of the game involves a combination of high and lower intensity bouts of activity and poses high physiological demands (Smith, 1998). More specifically, intense movements last between 7 and 14 seconds (Hohmann and Frase, 1992) and players' heart rate exceeds 80% of maximum for the majority of the game (Hollander et al., 1994). However, recent implementation of new rules has changed the duration of playing and rest, thus potentially altering the demands of the game.

Physiological measurements can provide an indication of improved athletic performance, serve as selection criteria (Gabbett, 2002a), performance predictors (Ugarkovic et al., 2002) or assisting in identifying the demands of a sport and/or a position (Babic et al., 2001). Testing specific to the sport, ideally conducted in the athlete's training environment, can obtain valid and reliable results closely related to the demands of a sport. Anthropometric measurements have also been widely used to assist in player development and training goals (Dowson et al., 2002), in establishment of somatotypes for certain sports and/or positions (Gabbett, 2002b) and talent identification (MacDougal et al., 1991).

A number of studies on male water polo players have investigated physiological and anthropometric characteristics (Csende et al., 1998; Aziz et al., 2002; Frenkl et al., 2001; Mazza et al., 1994; Carter and Marfell-Jones, 1994). Results from these studies are shown in Table 11.1.

Table 11.1 Results from previous studies conducted on elite male water polo players.
Results reported as mean ± sd.

Study	Sample	Height (cm)	Mass (kg)
Mazza *et al.* (1994)	Elite players	186.5 ± 6.5	86.1 ± 8.4
Csende *et al.* (1998)	Hungarian National team players	184.5 ± 9.1	87.4 ± 6.3
Frenkl *et al.* (2001)	Hungarian National team players	190.9 ± 6.0	91.1 ± 7.7
Aziz *et al.* (2002)	Singapore National team	178.5 ± 3.9	71.0 ± 8.4

These results indicate a variation in anthropometric characteristics between players of different countries but also between different assessment years. Unfortunately, due to different formats of data reporting, it was impossible to draw comparisons between the studies for the physiological characteristics. The development of a comprehensive battery of tests that includes both anthropometric and physiological measurements is thus needed (Tsekouras *et al.*, 2005).

However, there is only one study that has investigated female water polo players' characteristics. Carter and Marfell-Jones (1994) measured the anthropometric characteristics of female players but, to date, no study has investigated physiological characteristics alone or in combination with anthropometry. The study by Carter and Marfell-Jones (1994) assessed players competing at the 1991 World Championships, so there is no information available on the anthropometric characteristics of female water polo players during the last 15 years. The need for updated measurements is clear (Johnson *et al.*, 1989), as it is possible that during the intervening 15 years, the anthropometric characteristics of female water polo players will have been altered. Possible reasons are, (a) different training regimens due to advances in training methods, (b) modifications in the regulations of the game (resulting, subsequently, in changes in training programmes and tactics) and, (c) specific morphological adaptations due to the above.

A situation where physiological, anthropometric and sport-specific data can be obtained simultaneously provides the most accurate and informative results due to the ease of comparisons and the complete profiling achieved (Rodríguez, 1994; Tumilty *et al.*, 2000). Thus, the aim of this study was to determine the anthropometric and physiological characteristics of elite female water polo players.

2 METHODS

2.1 Subjects

The subjects were fourteen female players from the Scottish National team (mean ± sd: age 22 ± 4.4 years, height 168.7 ± 7.9 cm, body mass 65.9 ± 6.1 kg) who competed at the 2006 Commonwealth Water Polo Championship in Perth, Australia. All the players have been part of the squad for a minimum of 3 years. The study was approved by the Edge Hill University Ethics Committee and the players provided written, informed consent to participate.

2.2 Procedures

Testing took place over a four-day period during the team's preparatory training phase. The battery of tests utilised was based on selected anthropometrical and physiological characteristics. This comprised both laboratory and sport-specific protocols. All subjects were familiarised with the procedures prior to testing. The subjects had been instructed to refrain from strenuous exercise for forty-eight hours prior to testing and to avoid food and caffeine intake for two hours preceding the assessments.

The order of the tests was anthropometric measurements, aerobic power, anaerobic power, strength and flexibility in order to minimise the effects of previous tests on subsequent test performance, as suggested by the American College of Sports Medicine (1995). The equipment was calibrated according to manufacturers' standardised procedures.

2.2.1 Anthropometric measurements

For all the anthropometric measurements, standard International Society for the Advancement of Kinanthropometry (ISAK) procedures were followed. Stretch stature (height) was measured to the nearest 0.1 cm using a stadiometer (Harpenden, UK). Body mass was measured using a calibrated balance beam (Seca, UK) scale and was recorded to the nearest 0.1 kg. Body Mass Index (BMI) was calculated as $mass(kg)/height^2(m)$. Sum of four skinfolds (SUM4SF) was calculated from measurements at the biceps, triceps, subscapular and suprailiac sites using skinfold callipers (Harpenden, UK). The measurements were taken to the nearest 1 mm. Somatotype was calculated using the Heath-Carter protocol (Carter and Heath, 1990).

2.2.2. Aerobic power

Aerobic power was assessed using an incremental treadmill (Powerjog, UK) test to exhaustion. Initially, a 5-minute warm up was conducted at a velocity of 9 $km \cdot h^{-1}$. The test started at 10 $km \cdot h^{-1}$ and velocity was increased by 1 $km \cdot h^{-1}$ every two minutes until the subject reached volitional exhaustion. A gradient of 1% was used for the duration of the test, as suggested by Jones and Doust (1996). Expired air was collected and analysed using a Cardio2 automated gas analyser (Medgraphics, UK) throughout the test, to determine peak oxygen uptake. Heart rate was recorded by using a telemetric heart rate monitor (Polar, Finland). At the end of the test, a capillary blood sample was taken from the earlobe and the blood sample was analysed for lactate concentration (Analox Instruments Ltd., UK). The treadmill test was utilised since a direct measurement of aerobic power was deemed important, as it allows a common point of comparison with other sporting groups. However, equipment restraints did not allow for a direct measurement of aerobic power using a more specific swim-based test.

Sports specific tests involved the Multistage Swimming Shuttle Test (MSST). During this continuous incremental test, the subjects had to swim a 10-m distance at a progressively increasing speed, starting at 0.9 m·s^{-1} and increasing by 0.05 m·s^{-1} every stage. Each stage lasted approximately one minute and the shuttles were signalled by an audio cue. Subjects continued until volitional exhaustion (Rechichi *et al.*, 2000). At the end of the test, peak blood lactate was assessed using the methods outlined above.

2.2.3. Anaerobic power

Anaerobic power was measured via the Wingate Anaerobic Test. Initially, a 5-min warm up was conducted at a workload of 100 W with a 5-sec sprint at 3 min, followed by a 5-min rest (Winter and MacLaren, 2001). The test required the subject to cycle maximally on an ergometer (Monarch, UK) for 30 seconds against a resistance of 7.5% body mass (Bar-Or *et al.*, 1977; Zając *et al.*, 1999). Peak power, mean power and fatigue index were calculated (Cranlea, UK). Additionally, leg power was estimated using a vertical jump test (Takei Scientific Inst. Co. Ltd, UK) to allow comparisons with other sports data. The subjects placed their hands on their hips and did not use any assisting arm movement during the jump, thus isolating the contribution of the lower limbs (Lees and Fahmi, 1994).

Sports-specific tests involved 14 repetitions of 25 m lengths, swum at maximal velocity every 30 seconds. The fastest 25 m length indicates the swim-specific anaerobic alactic capacity whilst the slope of the linear regression equation is an indication of the players' swim specific speed-endurance (Rodríguez, 1994). A 30-sec cross-bar jump test was performed, which is commonly used in water polo. During this test, the subjects had to repeatedly jump out of the water and touch the vertical bar of the water polo goal using breaststroke kicks, aiming to achieve as many jumps as possible in 30 sec.

2.2.4. Strength

Maximum leg, back and handgrip strength (left and right arm) were assessed using an appropriate dynamometer (Takei Scientific Inst. Co. Ltd). The subjects had to adopt a squat position for the leg strength tests and a stiff leg dead lift position for the back strength test. Once in position, the subjects had to perform a maximal contraction. For the handgrip strength, the subject performed a maximal contraction while lowering an extended arm. The movement remained free from hitting external objects or the body and no extraneous movement was allowed. The best result out of three efforts was recorded for all tests.

2.2.5. Flexibility

Flexibility of the lower back and hip was determined by the sit and reach test. Subjects were required to sit with their heels placed against the edge of the sit and reach box, with the legs extended and to reach forward as far as possible in a controlled manner. During the reach, the subjects were instructed to exhale and maintain their head lowered between the arms. The best result out of three efforts was recorded.

Shoulder flexibility was measured using videography and commercially available software to measure angles (Quintic, UK). Body landmarks on the right arm (top of humerus and olecranon) were marked. The subjects hyperextended their arm along the sagittal plane, with their upper body upright. No rotation of the upper body was allowed. Shoulder hyperextension angle was measured relative to the vertical line (vertical line = $0°$).

2.3 Statistical analysis

Descriptive statistics were calculated for all variables and Pearson's correlation (r) used to test for significant relationships. Significance level was set at $P < 0.05$. All statistical analyses were conducted using SPSSv13.0.

3 RESULTS

Descriptives (mean ± sd) of the results can be found in Table 11.2.

Table 11.2 Anthropometric and physiological assessment results.

Test	Results
Height (cm)	168.7 ± 7.9
Mass (kg)	65.9 ± 6.1
BMI	23.3 ± 2.5
Skinfolds (mm)	
- SUM4SF	37.4 ± 10.6
- biceps	6.6 ± 2.4
- triceps	11.6 ± 2.8
- subscapular	11.3 ± 3.4
- suprailiac	8.0 ± 5.5
Somatotype	
- endomorphy	3.3 ± 1.1
- mesomorphy	5.1 ± 1.5
- ectomorphy	1.9 ± 1.2
Peak oxygen uptake (L·min^{-1})	3.39 ± 0.37
Peak oxygen uptake (ml·kg^{-1}·min^{-1})	51.4 ± 4.5
Peak blood lactate (mmol·L^{-1}) (treadmill)	6.3 ± 1.4
Peak heart rate (bpm)	196 ± 9
MSST (shuttles)	33.0 ± 8.7
Peak blood lactate (mmol·L^{-1}) (MSST)	6.8 ± 1.7
Peak power (W)	667.8 ± 93.7
Mean power (W)	459.2 ± 45.3

Fatigue index (%)	48.3 ± 7.1
Vertical jump (cm)	33.4 ± 3.3
14 x 25 m	
- maximum speed (m·s^{-1})	1.51 ± 0.07
- speed gradient	−0.0061 ± 0.0072
Cross-bar jumps (jumps)	22 ± 2
Leg strength (kg)	118.9 ± 30.8
Back strength (kg)	84.8 ± 11.7
Handgrip strength (kg)	
- left arm	30.9 ± 3.6
- right arm	33.7 ± 3.9
Flexibility	
- Sit and reach (cm)	27 ± 9.7
- Shoulder hyperextension (°)	20 ± 8

For all variables examined, significant correlations were found for SUM4SF with leg strength. Furthermore, peak oxygen uptake (ml·kg^{-1}·min^{-1}) was significantly correlated with fatigue index, back strength, leg strength and MSST. No other correlations were found to be significant. Correlation coefficients and significance for variables significantly related to SUM4SF and peak oxygen uptake are presented in Table 11.3.

Table 11.3 Pearson's correlation (*r*) for significant relationships (p < 0.05) between variables examined.

	SUM4SF	Peak oxygen uptake			
	Leg strength	Fatigue Index	Back strength	Leg strength	MSST
R	−0.595	−0.650	0.647	0.658	0.660
Sig.	0.041	0.021	0.023	0.015	0.014

4 DISCUSSION

The Scottish National water polo female team members were shorter (168.7 ± 7.9 cm) than the elite female water polo players studied by Mazza *et al.* (1994) and Rechichi *et al.* (2000) who reported heights of 171.3 ± 5.9 cm and 173.1 ± 4.5 cm, respectively. Tsekouras *et al.* (2005) postulated that body size plays an important role in the game of water polo, with taller players having an advantage in reaching and controlling passes. However, there was no trend with respect to body mass, as the Scottish team were heavier (body mass 65.9 ± 6.1 kg) than 64.8 ± 7.2 kg reported by Mazza *et al.* (1994) but lighter than 70.6 ± 8.9 identified by Rechichi *et al.* (2000). Sum of various skinfold sites has been reported but there has been no consistent trend in the skinfold sites used in previous studies. For example, Carter and Marfell-Jones (1994) used six skinfold sites while Aziz *et al.* (2002) used three. However, comparison of individual skinfold sites (biceps 6.6 ± 2.4 mm, triceps 11.6 ± 2.8 mm, subscapular 11.3 ± 3.4 mm and suprailiac 8.0 ± 5.5 mm) showed the current findings to be similar to the values (7.1 ± 2.7 mm, 15.3 ± 4.3 mm, 10.5 ± 3.6 mm, 9.6 ± 3.4 mm for biceps, triceps, subscapular and suprailiac, respectively) reported by Drinkwater and Mazza (1994). The mean somatotype in the current study was found to be predominantly mesoendomorphic. Carter and Marfell-Jones (1994) found that the somatotype of elite female water polo players was 3.5-4-3, which suggests that the musculature of the players in the current study was higher. This morphological difference could reflect the need for greater muscularity due to the evolving nature of the game.

Physiological assessment showed aerobic power was comparable to other invasive games players such as the English Women's Soccer team (Davis and Brewer, 1992). Peak oxygen uptake was slightly higher than that recorded by Rechichi *et al.* (2000) in female water polo players. However, the current study assessed aerobic power using a treadmill-based test whilst Rechichi *et al.* (2000) utilised a tethered swimming protocol and therefore direct comparisons of values are not possible. Peak heart rate was also higher than that recorded by Rechichi *et al.* (2000). This supports the findings of DiCarlo *et al.* (1991) who found a significant difference in peak heart rates during maximal running and swimming, possibly due to differences in muscle mass employed, the horizontal position in swimming which aids venous return and increases stroke volume and the fact that heat dissipation may be facilitated in cool water. The peak lactate recorded at the end of the treadmill-based test was lower than the 8 mmol·L^{-1} anticipated at the end of maximal exercise, although slightly higher than the values recorded by Konstantaki *et al.* (1998) at the end of swim bench exercise. However, heart rate, rating of perceived exertion and respiratory exchange ratio indicated maximal effort had been attained.

The MSST is a more sports-specific test and has been shown to be a reliable and valid test for elite water polo players (Rechichi *et al.*, 2000). MSST scores achieved in the current study were lower compared to the 40.9 ± 11.7 shuttles reported by Rechichi *et al.* (2000) for elite female water polo players, as was peak lactate (8.9 ± 2.5 mmol·L^{-1}). A highly developed aerobic system is vital to assist in recovering from the high intensity bouts and delaying the onset of fatigue, thus it is suggested that the sports-specific aerobic power of the subjects in the current study should be improved via appropriate training.

Peak power achieved during the Wingate anaerobic test was deemed excellent, according to norms produced by Maud and Schultz (1989). This is beneficial to the water polo player since sprinting bouts during the game tend to last less than 10 seconds (Smith, 1991), thus relying on alactic energy metabolism. However, those norms were based on active rather than elite athletic population and the comparisons need to be interpreted with caution. Perhaps a more indicative measure of the subjects' performance was the mean power and the fatigue index. The fatigue index of 48.3% in combination with the relatively low mean power (68.8% of peak power) indicates the inability of the subjects to resist fatigue. The need to have a low fatigue index is essential, as the demands of the game require complex and alternating high power activities.

Vertical jump results rated poorly with norms reported by Chu (1996). However, the norms were based on elite athletes whose sports required them to exert high forces against the ground. The nature of water polo necessitates players to produce high upward forces less frequently compared to the above athletes but also to generate them while pushing against water rather than a fixed resistance. Although the specific demands of water polo partially account for this low score, given the peak power results discussed earlier, higher vertical jump scores would be expected.

The 14 x 25 m shuttle test is a better indication of some of the anaerobic abilities of the players. Unfortunately, there are no published data for females to which the present results could be compared. Rodríguez (1994) reported mean velocities of 1.83 $m·s^{-1}$ for elite male Spanish water polo players. However, it is the authors' view that the maximum speed of 1.51 $m·s^{-1}$ is moderate for the playing level of the female subjects. Nevertheless, the surprisingly small drop of speed throughout the shuttles may partially explain the relatively lower speed. Despite the instructions and encouragement given, it was possible for the athletes to have paced themselves through the 14 repetitions, achieving a lower maximum speed with a corresponding small gradient decrease. It is suggested that a 25 m sprint performed on another occasion would enable comparisons with the maximum speed achieved during this test.

In the cross-bar jumps, the subjects scored moderately low. It needs to be noted that the comparison is based on anecdotal evidence and it is only reported in order to construct a complete profile of the team for future studies. The validity and reliability of the test remains to be established in order to allow for more accurate comparisons (Bampouras and Marrin, 2006).

A successful shot or pass relies on the action of the lower body, creating an upwards impulse to achieve optimal height out of the water (Sanders, 1999a; Davis and Blanksby, 1977) as well as arm strength (Bloomfield *et al.*, 1990). The kinetic chain of this action involves a number of muscle groups and more importantly, the lower limbs (Sanders, 1999b) and the back muscles (Platanou, 2005) to drive the body upwards and the arms to push the ball forwards. The results for the leg strength test indicated that the subjects ranked as 'good' whilst the results for the back, left and right handgrip strength ranked as 'average' (Heyward, 1997) compared to non-elite athletes. Given the importance of the leg, back and arm muscles in execution of water polo skills, it is suggested that higher values would be beneficial.

The eggbeater action used in water polo is a complex movement, requiring high levels of flexibility (Sanders, 1999a,b). The results for the sit and reach test revealed that the subjects had good flexibility compared to the non-elite population (Fitness Canada, 1986). However, care needs to be taken in the interpretation of the results due to the different activity demands between elite water polo players and physically active populations. Shoulder hyperextension results are similar to findings by Morales and Arellano (2006) on young, elite swimmers. Given the younger age of swimmers in Morales and Arellano's study compared to the current study as well as the different demands game activities place on the water polo player's shoulder, the flexibility of the subjects is deemed good. However, care needs to be given to prevent the shoulder joint becoming hypermobile as this could result in shoulder impingement, a common problem amongst swimmers (Kenal and Knapp, 1996).

SUM4SF was negatively correlated with leg strength. Passing, shooting or blocking are important elements of the game and require high levels of lower limb strength (Sanders, 1999b). Therefore, it could be postulated that a high level of body fat would be a disadvantage to the water polo player in these game situations. Although it has been suggested that the adiposity of water polo players may represent a physical advantage in terms of buoyancy (Drinkwater and Mazza, 1994), this advantage could be minimised by the effort required to overcome greater moment of inertia during accelerations in the game (Tumilty *et al.*, 2000). Hence, a lower body fat level would be beneficial to players in the current study.

Additionally, peak oxygen uptake was significantly correlated with fatigue index, back strength, leg strength and MSST with moderate correlations. Although, as discussed earlier, fatigue index is representative of the ability of an individual to resist fatigue during the Wingate anaerobic test, it has been established by Beneke *et al.* (2002) that the percentage contributions of energy from aerobic, anaerobic alactic and lactic acid metabolism was $18.6 \pm 2.5\%$, $31.1 \pm 4.6\%$, and $50.3 \pm 5.1\%$, respectively, thus the contribution of the aerobic energy system can be deemed significant. Hence, a higher aerobic power could result in a lower fatigue index. MSST was found to be significantly and moderately correlated with peak oxygen uptake. Rechichi *et al.* (2000) reported a high ($r = 0.854$) and significant correlation of MSST to peak oxygen uptake. Whilst the present correlation was lower, it should be reinforced that Rechichi *et al.* (2000) assessed peak oxygen uptake using tethered swimming. Thus, it would be expected that two swim-based tests would present higher agreement compared to a swim-based and a running-based test. Water polo is unique in the demands it places on the athlete, both aerobic and anaerobic (Andreoli *et al.*, 2004), which dictates concurrent strength and endurance training to be utilised. The correlation of peak oxygen uptake with leg and back strength may appear spurious. However, although the players are members of a National team, they are heterogeneous in terms of age and experience and, hence, there may be differences in the volume of training between team members.

The present study fills an acknowledged gap in literature with reference to female water polo anthropometric and physiological characteristics. The last studies were conducted during the 1991 World Championships (Mazza *et al.*, 1994; Carter and Marfell-Jones, 1994). Since then, no studies have specifically assessed these characteristics in female players. The present study is the first to have provided a profile of elite female water polo players with reference to both anthropometric and physiological characteristics. It utilised both laboratory and sport-specific tests providing information for future assessment, comparison and monitoring. This is important to the sports scientist in relation to changes and adaptations of anthropometric and physiological characteristics and to the coach and athlete with regards to the effectiveness and appropriateness of the training programme, the athlete's physical condition and/or preparation for competition (Johnson *et al.*, 1989) as well as the positional differences (Pavicic *et al.*, 2000). Furthermore, the changing demands of the game would impact on the training regimens used and, therefore, dictate the need for regular monitoring over prolonged periods of time. Lozovina and Pavicic (2004) in their 15-year survey, found significant morphological adaptations in male water polo players whilst the present study suggests that over the intervening years the players have become more muscular. More longitudinal studies are required: (a) to verify the findings of this study (Subotnik and Arnold, 1993); (b) to investigate changes over time; and (c) for talent identification purposes.

Peak oxygen uptake is the strongest correlate to other physiological and anthropometric variables, as demonstrated by the current findings. It is suggested that training aims to maximise aerobic power and minimise unnecessary body fat levels. Additionally, increased muscularity of the players combined with the respective demands of the sport, renders appropriate strength training necessary. Consequently, coaches and athletes should ensure that aerobic endurance training is combined with resistance training to maximise performance (Leveritt *et al.*, 1999). The introduction of new rules, extending the duration of the game has altered the metabolic demands. Therefore, the above recommendations become important for successful performance in water polo.

The current study used fourteen female water polo players, which constitutes a small sample and does not permit the examination of positional differences. Although it was attempted to utilise a range of tests reflecting the demands of the game, as previously mentioned, equipment restraints did not allow for a direct measurement of aerobic power or lactate threshold using a swim-based test. Finally, whilst testing took place in the preparatory phase, the training load of the athletes was not made available. Further studies should incorporate a larger sample examining all positions and considering the training load.

5 CONCLUSION

A profile of the anthropometric and physiological characteristics of an elite female water polo team was provided. This information can be utilised by scientists, coaches and athletes for assessment, comparison and monitoring of training and performance. The Scottish national female water polo team demonstrated higher muscularity compared to elite female water polo players in 1991. Peak oxygen uptake was relatively high and was found to be the strongest correlate to other variables. It was suggested that water polo training should maximise aerobic power and concurrently utilise resistance training to ensure that performance matches the changing demands of the game.

6 REFERENCES

American College of Sports Medicine, 1995, *ACSM's Guidelines for Exercise Testing and Prescription* (Baltimore: Williams and Wilkins).
Andreoli, A., Melchiorri, G., Volpe, S.L., Sardella, F., Iacopino, L. and De Lorenzo, A., 2004, Multicompartmental model to assess body composition in professional water polo players. *Journal of Sports Medicine and Physical Fitness*, **44**, 38–43.
Aziz, A.R., Lee, H.C. and Teh, K.C., 2002, Physiological characteristics of Singapore national water polo team players. *Journal of Sports Medicine and Physical Fitness*, **42**, 315–319.

Babic, Z., Misigoj-Durakovic, M., Matasic, H. and Jancic, J., 2001, Croatian rugby project-part I: Anthropometric characteristics, body composition and constitution. *Journal of Sports Medicine and Physical Fitness*, **41**, 250–256.

Bampouras, T.M. and Marrin, K., 2006, Validity and reliability of a commonly used water polo test: A pilot study. *Portuguese Journal of Sport Sciences*, **6**, 72–73.

Bar-Or, O., Dotan, R. and Inbar, O., 1977, A 30-seconds all-out ergometric test – its reliability and validity for anaerobic capacity. *Israel Journal of Medical Sciences*, **13**, 126.

Beneke, R., Pollmann, C., Bleif, I., Leithauser, R.M., and Hutler, M., 2002, How anaerobic is the Wingate Anaerobic Test for humans? *European Journal of Applied Physiology*, **87**, 388–392.

Bloomfield, J., Blanksby, B.A., Ackland, T.R., and Allison, G.T., 1990, The influence of strength training on overhead throwing velocity of elite water polo players. *Australian Journal of Science and Medicine in Sports*, **22**, 63–67.

Carter, J.E.L. and Heath, B.H., 1990, *Somatotyping-development and Applications* (Cambridge: Cambridge University Press).

Carter, J.E.L. and Marfell-Jones, M.J., 1994, Somatotypes. In *Kinanthropometry in Aquatic Sports*, edited by Carter, J.E.L. and Ackland, T.R. (Champaign: Human Kinetics), pp. 55–82.

Chu, D.A., 1996, *Explosive Power and Strength* (Champaign: Human Kinetics).

Csende, Z., Mészáros, J., Tihanyi, J. and Zsidegh, M., 1998, Body composition and cardiorespiratory characteristics of world class waterpolo and kayak athletes. *Coaching and Sport Science Journal*, **3**, 9–13.

Davis, J.A. and Brewer, J., 1992, Physiological characteristics of an international female soccer squad. *Journal of Sports Sciences*, **10**, 142–143.

Davis, T., and Blanksby, B.A., 1977, Cinematographical analysis of the overhand water polo throw. *Journal of Sports Medicine*, **17**, 5–16.

DiCarlo, L.J., Sparling, P.B., Millard-Stafford, M.L., and Rupp, J.C., 1991, Peak heart rates during maximal running and swimming: Implications for exercise prescription. *International Journal of Sports Medicine*, **12**, 309–312.

Dowson, M.N., Cronin, J.B. and Presland, J.D., 2002, Anthropometric and physiological differences between gender and age groups of New Zealand national soccer players. In *Science and Football IV*, edited by Spinks, W. (London: Routledge), pp. 63–71.

Drinkwater, D.T. and Mazza, J.C., 1994, Body Composition. In *Kinanthropometry in Aquatic Sports*, edited by Carter, J.E.L. and Ackland, T.R. (Champaign: Human Kinetics), pp. 102–137.

Fitness Canada, 1986, *Canadian Standardised Test of Fitness Operations Manual*, 3rd ed. (Ottawa: Minister of State, Fitness and Amateur Sport).

Frenkl, R., Mészáros, J., Soliman, Y.A., and Mohácsi, J., 2001, Body composition and peak aerobic power in male international level Hungarian athletes. *Acta Physiologica Hungarica*, **88**, 251–258.

Gabbett, T.J., 2002a, Influence of physiological characteristics on selection in a semi-professional first grade rugby league team: a case study. *Journal of Sports Sciences*, **20**, 399–405.

Gabbett, T.J., 2002b, Physiological characteristics of junior and senior rugby league players. *British Journal of Sports Medicine*, **36**, 334–340.

Grantham, N.J., 2000, Body composition and physiological characteristics of male and female national and international high-performance gymnasts. *Journal of Sport Sciences*, **18**, 24.

Heyward, V.H., 1997, *Advanced Fitness Assessment and Exercise Prescription*, 3rd ed. (Champaign: Human Kinetics).

Hohmann, A. and Frase, R., 1992, Analysis of swimming speed and energy metabolism in competition water polo games. In *Swimming Science VI: Biomechanics and Medicine in Swimming*, edited by MacLaren, D., Reilly, T., and Lees, A. (London: E & F Spon), pp. 313–319.

Hollander, A.P., Dupont, S.H.G. and Volkerigk, S.M., 1994, Physiological strain during competitive water polo games and training. In *Medicine and Science in Aquatic Sports*, edited by Miyashita, M., Mutoh, Y., and Richardson, A.D. (Basel: Karger), pp. 178–185.

Johnson, G.O., Nebelsick-Gullett, L.J., Thorland, W.G. and Housh, T.J., 1989, The effect of a competitive season on the body composition of university female athletes. *Journal of Sports Medicine and Physical Fitness*, **29**, 314–320.

Jones, A. and Doust, J., 1996, A 1% treadmill grade most accurately reflects the energetic cost of outdoor running. *Journal of Sports Sciences*, **14**, 321–327.

Kenal, K.A. and Knapp, L.D., 1996, Rehabilitation of injuries in competitive swimmers. *Sports Medicine*, **22**, 337–347.

Konstantaki, M., Trowbridge, E.A., and Swaine, I.L., 1998, The relationship between blood lactate and heart rate responses to swim bench exercise and women's competitive water polo. *Journal of Sports Sciences*, **16**, 251–256.

Lees, A. and Fahmi, E., 1994, Optimal drop heights for plyometric training. *Ergonomics*, **37**, 141–148.

Leveritt, M., Abernethy, P.J., Barry, B.K. and Logan, P.A., 1999, Concurrent strength and endurance training: a review. *Sports Medicine*, **28**, 413–427.

Lozovina, V. and Pavicic, L., 2004, Anthropometric changes in elite male water polo players: survey in 1980 and 1995. *Croatian Medical Journal*, **45**, 202–205.

MacDougal, J.D., Wenger, H.A. and Green, H.J., 1991, *Physiological Testing of the High Performance Athlete* (Leeds: Human Kinetics).

Maud, P.J. and Schultz, B.B., 1989, Norms for the Wingate anaerobic test with comparison to another similar test. *Research Quarterly for Exercise and Sport*, **60**, 144–151.

Mazza, J.C., Ackland, T.R., Bach, T.M. and Cosolito, P., 1994, Absolute body size. In *Kinanthropometry in Aquatic Sports*, edited by Carter, J.E.L. and Ackland, T.R. (Champaign: Human Kinetics), pp. 15–54.

Morales, E. and Arellano, R., 2006, A three-year follow up study of age group swimmers: anthropometric, flexibility and counter-movement jump force recordings. *Portuguese Journal of Sports Sciences*, **6**, 307–309.

Pavicic, L., Tomany, E. and Lozovina, V., 2000, A study of anthropometric differences among elite water polo players with different team role assignments. In *Proceedings of the 2000 Pre-Olympic Congress*, Brisbane (Brisbane: Australian Sports Commission), p. 503.

Platanou, T., 2005, On-water and dryland vertical jump in water polo players. *Journal of Sports Medicine and Physical Fitness*, **45**, 26–31.

Rechichi, C., Dawson, B., and Lawrence, S.R., 2000, A multistage shuttle swim test to assess aerobic fitness in competitive water polo players. *Journal of Science and Medicine in Sport*, **3**, 55–64.

Rodríguez, F.A., 1994, Cardiorespiratory and metabolic field testing in swimming and water polo: from physiological concepts to practical methods. In *VIII International Symposium on Biomechanics and Medicine in Swimming*, Jyväskylä, edited by Keskinen, K.L., Komi, P.V. and Hollander, A.P. (Jyväskylä: Gummerus Printing), pp. 219–226.

Sanders, R.H, 1999a, Analysis of the eggbeater kick used to maintain height in water polo. *Journal of Applied Biomechanics*, **15**, 284–291.

Sanders, R.H., 1999b, A model of kinematic variables determining height achieved in water polo 'boosts'. *Journal of Applied Biomechanics*, **15**, 270–283.

Smith, H.K. 1991, Physiological fitness and energy demands of water polo: time-motion analysis of field players and goaltenders. In *Proceedings of the Federation Internationale de Natation Amateur (FINA) First World Water Polo Coaches Seminar*, Athens (Lausanne: FINA), pp. 183–207.

Smith, H.K., 1998, Applied physiology of water polo. *Sports Medicine*, **26**, 317–334.

Subotnik, R.F. and Arnold, K.D., 1993, Longitudinal studies of giftedness: Investigating the fulfilment of promise. In *International Handbook of Research and Development of Giftedness and Talent*, edited by Heller, K.A., Monks, F.K. and Passow, A.H. (New York: Pergamon), pp. 149–160.

Tsekouras, Y.E., Kavouras, S.A., Campagna, A., Kotsis, Y.P., Syntosi, S.S., Papazoglou, K. and Sidossis, L.S., 2005, The anthropometrical and physiological characteristics of elite water polo players. *European Journal of Applied Physiology*, **95**, 35–41.

Tumilty, D., Logan, P., Clews, W. and Cameron, D., 2000, Protocols for the physiological assessment of elite water polo players. In *Physiological Tests for Elite Athletes*, edited by Gore I. (Champaign: Human Kinetics), pp. 411–421.

Ugarkovic, D., Matavulj, D., Kukolj, M. and Jaric, S., 2002, Standard anthropometric, body composition, and strength variables as predictors of jumping performance in elite junior athletes. *Journal of Strength and Conditioning Research*, **16**, 227–230.

Winter, E.M. and MacLaren, D.P., 2001, Assessment of maximal intensity exercise. In *Kinanthropometry and Exercise Physiology Laboratory Manual: Tests, Procedures and Data*, edited by Eston, R. and Reilly, T. (Glasgow: Routledge), pp. 263–288.

Zając, A., Jarząbek, R. and Waśkiewicz, Z., 1999, The diagnostic value of the 10- and 30-second Wingate test for competitive athletes, *Journal of Strength and Conditioning Research*, **13**, 16–19.

CHAPTER TWELVE

Kinanthropometric differences between playing levels and position in Rugby Union

T.L.A. Doyle[1], J.W.L. Keogh[2] and J. Presland[3]

[1] School of Human Movement and Exercise Science, University of Western Australia, Western Australia, Australia

[2] Institute of Sport and Recreation Research New Zealand, Division of Sport and Recreation, Auckland University of Technology, Auckland, New Zealand

[3] Black Ferns Fitness Trainer, New Zealand Rugby Union, New Zealand

1 INTRODUCTION

It is common practice for professional sports to use talent identification to highlight players who may have potential for future success in a given sport. For example, in Australia the Australian Football League (AFL) run a draft camp every year where young hopefuls are assessed on anthropometric measures as well as performance-related measures. Similarly, in the USA the National Football League runs a combined camp where testing is conducted for the same purpose as the AFL draft camp. For such camps to be effective it is necessary to know what characteristics are important in the elite performers of the sport. Further, in preparing for that sport, whether it be preparation for the draft process or for game play itself, knowledge about the physical and performance standards of elite players is useful to set goals and to help program training schedules.

Investigations in rugby league have determined that higher level players are superior to lower level players across a number of anthropometric and performance-related measures (Baker, 2001, 2002). These studies report that professional rugby league players were stronger and more powerful in upper body measures than lower level players. These differences existed even though body mass and height did not differ significantly between the player groups. Though a good proportion of these strength differences may be the result of many years of training, it was suggested that younger developing players aim to attain the strength and power levels of their more senior counterparts. In yet another code of football, some differences in physiological measures existed in a group of professional australian rules football players when comparing starters and non-starters (Young *et al.*, 2005). Height and body mass were not significantly different between these two groups, however starters were faster over 10 m than non-starters together with greater jumping ability and aerobic power as measured with an intermittent recovery test similar to a shuttle run (yo-yo test). The tests used to describe the profiles of players in these sports, must be specific to the requirements of the game played; consequently, it may be more appropriate to use a test battery specific to player position.

In the rugby union time-motion analysis literature, a typical rugby union match involves repeated short-duration high-intensity activities such as sprinting, jumping, tackling and scrummaging that are interspersed with longer-duration, lower-intensity activities such as jogging and walking (Deutsch *et al.*, 2002; Docherty *et al.*, 1988; Duthie *et al.*, 2005; James *et al.*, 2005). The attainment of relatively high to high levels of anthropometric and performance-related variables (e.g. muscle mass, sprinting speed, aerobic power and muscular strength) would appear to confer advantages in rugby union competition (Nicholas, 1997; Quarrie and Williams, 2002; Quarrie *et al.*, 1995). As a result, more-successful rugby union teams have been found to have a distinct anthropometric and performance-related profile when compared to less successful teams (Rigg and Reilly, 1988; Mayes and Nuttall, 1995).

It must however be acknowledged that the different positions on the rugby field have varying roles and that all of these roles contribute to the overall success of the team. The primary role of the forwards is to secure the ball in high-intensity contact situations such as scrums, lineouts, rucks and mauls and tackles, whereas the backs are more concerned with evading the opposition and scoring tries (Duthie *et al.*, 2003; James *et al.*, 2005). Consistent with these somewhat position-specific requirements and movement patterns inherent to the sport of rugby union, senior club (SC) forwards are typically taller, heavier, stronger, possess greater levels of muscle and fat mass and can produce greater total body momentum than SC backs (Quarrie *et al.*, 1995, 1996; Rigg and Reilly, 1988). However, these studies also indicate that SC forwards may have lower levels of relative muscular endurance, aerobic power and sprinting speed than the backs. It is however relatively unknown if these inter-positional differences in kinanthropometry also occur in rugby players of a more elite standard.

Therefore, this research had the purpose of measuring anthropometric and performance-related measures of rugby union players from three different playing standards (international (INT), senior club (SC), and junior state representative (JS)) across all positions in a rugby team. In addition to standard descriptive measures of body stature, speed and aerobic power were measured using standard field testing, and upper body strength was assessed using repetition maximum testing methods. It was hypothesised that higher standard players would be superior to the lower standard players and that regardless of playing level, forwards would be significantly taller, heavier, stockier, and have greater total body momentum than backs. It was also expected that a greater number of inter-positional differences would be observed for senior club rather than the international or junior state players.

2 METHODS

2.1 Participants

A sample of convenience was used for this research that included players from three different training squads. In total, results from sixty-four players were used in this investigation. Descriptive data for these three groups is provided in Table 12.1. International (INT) players represented their country at the 2003 World Cup Rugby, senior club (SC) players play in a local senior competition at either 1st or 2nd grade level, and the junior state (JS) squad represents elite players sourced from local junior club and schools competitions.

2.2 Procedures

Testing took place at different locations for each group, but was always performed according to appropriate standards and by personnel accredited by their respective institution for conducting anthropometric and physical performance testing. In the case of the INT and SC squads, the Sports and Exercise Science New Zealand standards were followed (Bishop and Hume, 2000). The JS group were tested using the protocol outlined by the Laboratory Standards Assistance Scheme (Australian Institute of Sport, 2004). The procedures used at each location were the same for all tests unless noted below. For all testing, any prior physical activity that would interfere with the reliability and validity of the results was limited.

Standing height and body mass: While barefoot, athletes stood against a wall mounted stadiometer and their height was recording according to the national standard protocols (Norton and Olds, 1996). Body mass was measured using calibrated digital scales. Body mass index (BMI) was calculated from these two measures, using the following equation (12.1):

$$BMI = \frac{BodyMass}{Height^2} \tag{12.1}$$

10 m sprint and momentum: Sprinting speed in all groups was recorded using a dual-beam system with a precision of 0.01 s (Speed Light Athletic Timing System, Lismore, Australia). Players began in a stationary position and began the test in their own time. Timing began when they broke the first light beam. No rocking or movement was allowed prior to the start. Each player was allowed three trials and the best trial was used for analysis. Multiplying the player's average velocity by their body mass (measured near-nude as above) provided their average momentum over the 10 m distance (Quarrie *et al.*, 1996).

Aerobic power: The aerobic power of the players was measured using the multi-stage fitness test (MSFT). This test has been previously shown to be a valid measure (Grant *et al.*, 1995; Leger and Gadoury, 1989) and can be used to reliably estimate $\dot{V}O_2max$ (Leger *et al.*, 1988; Sproule *et al.*, 1993). The MSFT requires players to run back and forth over a 20 m distance (or shuttle) for as long as possible. After set periods, the time they have to make the 20 m distance decreases and when they are no longer able to complete two consecutive shuttles their test is ceased and their score recorded. From the raw MSFT score a $\dot{V}O_{2max}$ was estimated (Australian Coaching Council, 1988).

Upper body strength: Using a 1-repetition maximum (1RM) bench press test, participants' upper body strength was measured. The technique used for this test has been described elsewhere (Australian Institute of Sport, 2004). In summary the technique requires the lift to be controlled at all times, the weight cannot be 'bounced' off the chest, buttocks and shoulders must remain in contact with bench, and feet must remain in contact with the floor at all times. SC and JS groups performed a 3-repetition maximum (3RM) in accordance with testing protocol at the institution where the testing was conducted. In this instance, the 1RM was estimated from the 3RM result according to Baker (2004). Relative bench press strength was calculated using the Wilks score. This involved multiplying each player's 1RM by the Wilks ratio specific to their body mass. The Wilks score has been validated by Vanderburgh and Batterham (1999) and is used in all International Powerlifting Federation events to determine the Champion of Champions.

2.3 Statistical analysis

Results are provided as mean ± standard deviation. A multivariate analysis of variance was used to determine significant differences between groups the p-value as set at 0.05 as the criterion for determining a significant difference. Gabriel's post-hoc test was used to assess where group differences existed. Statistical analyses were performed using SPSS version 11.5 (SPSS Inc., Chicago, Il, USA).

3 RESULTS

3.1 Playing level

3.1.1 Descriptive data (Table 12.1)

Both the INT and SC groups were heavier than the JS group by 13–18 kg ($p < 0.05$). In each group the standard deviations were quite high, of the order of 10–15 kg. All groups were not statistically different for height, although the JS group were on average 3 cm shorter than the INT and SC groups. Following the same pattern as body mass the JS group were of slighter build than the INT and SC group according to the BMI.

Table 12.1 Descriptive data, mean(sd), for each of the groups used in the investigation. Significance is as indicated. Level of significance was set at $P < 0.05$.

	Age range (yrs)	Height (m)	Mass (kg)	BMI $(kg.m^{-2})$
INT ($n = 20$)	18–35	1.83(.06)	102.0(13.4)	30.3(3.1)
SC ($n = 25$)	18–35	1.83(.07)	97.3(15.0)	29.0(3.7)
JS ($n = 19$)	15–17	1.80(.07)	84.8(10.6)	26.2(2.1)
Sig. diff.	N/A	None	~§	~§

INT = International, SC = Senior Club, JS = Junior State
* = difference between INT and SC
~ = difference between SC and JS
§ = difference between JS and INT

3.1.2 Speed and aerobic power (Table 12.2)

All groups were statistically different from each other when compared for speed over 10 metres ($p < 0.05$). As expected, the INT group was the fastest and the JS squad the slowest. Related to speed, momentum followed this same pattern such that the INT had the most momentum over 10 m and the JS had the least, though there was no difference between the INT and SC groups. Interestingly, the only difference that was significant when comparing aerobic power based on the estimated $\dot{V}O_2$max from the MSFT, existed between the INT and SC groups. Though the JS group had similar results, the JS group was not determined to be significantly different from either of the other two groups.

Table 12.2 Results, mean(sd), from speed and aerobic power testing. Significance is as indicated. Level of significance was set at $P < 0.05$.

	10 m (s)	Momentum $(kg.m.s^{-1})$	$\dot{V}O_2$max $(ml.kg^{-1}.min^{-1})$
INT	1.67(.06)	606(74)	52.5(4.6)
SC	1.76(.06)	552(78)	47.7(6.2)
JS	1.87(.04)	453(56)	48.5(5.9)
Sig. diff.	*~§	~§	*

INT = International, SC = Senior Club, JS = Junior State
* = difference between INT and SC
~ = difference between SC and JS
§ = difference between JS and INT

3.1.3 Upper body strength (Table 12.3)

The 1RM scores, predicted from a 3RM score for the SC and JS groups, were significantly different between all groups ($p < 0.05$). Unsurprisingly, the JS group were the weakest (86 ± 15 kg) and the INT (133 ± 17 kg) were the strongest with the SC (105 ± 22) being intermediate. When the absolute measures were converted to relative measures based on body weight, all groups were significantly different from each other ($p < 0.05$).

Table 12.3 Results, mean(sd), for upper body strength testing. 1RM scores were estimated from a 3RM test for the SC and JS groups. Significance is as indicated. Level of significance was set at $P < 0.05$.

	1RM Bench Press (kg)	*Rel. Bench Press (Wilk's Score)*
INT	133(17)	1.32(.17)
SC	105(22)	1.08(.20)
JS	86(15)	1.01(.16)
Sig. diff.	* ~ §	* ~ §

INT = International, SC = Senior Club, JS = Junior State
* = difference between INT and SC
~ = difference between SC and JS
§ = difference between JS and INT

3.2 Positional differences

Tables 12.4–12.6 present the anthropometric characteristics of the INT, SC and JS players, respectively. In the INT group, forwards were significantly taller, heavier and stockier (as determined by the BMI) and had greater total body momentum over 10 m than the backs ($p < 0.05$). No significant inter-positional differences were observed for the INT players for 10 m sprint time, aerobic power or bench press strength (absolute strength).

Senior club forwards were significantly taller, heavier and stockier (as assessed by the BMI) than the SC backs and the forwards had significantly greater total body momentum ($p < 0.05$). Although absolute bench press strength and ten metre sprint times were significant at the $p < 0.10$ level, neither result reached the criterion level of $p < 0.05$. No significant inter-positional differences were found in the SC group for aerobic power or relative bench press strength.

In the JS group, the forwards and backs were not statistically different across all measures.

Table 12.4 Differences in the anthropometric and performance-related characteristics of international (INT) forwards and backs. Level of significance was set at $P < 0.05$.

	Forwards (n = 10)	Backs (n = 10)
Height (cm) *	186.4(4.8)	179.0(4.3)
Body mass (kg) *	110.5(11.1)	92.6(9.0)
Body mass Index (kg.m^{-2}) *	31.8(3.2)	28.9(2.2)
10 m sprint (s)	1.69(0.07)	1.66(0.05)
Average 10 m Momentum (kg.m.s^{-1}) *	653.8(60.6)	558.3(52.1)
Aerobic Power (ml.kg.min^{-1})	51.6(5.4)	53.5(3.7)
Absolute 1RM bench press (kg)	135.7(17.8)	130.2(15.7)
Rel. 1RM bench press (Wilks Score)	80.0(8.7)	82.4(9.5)

* indicates a significant difference between forwards and backs

Table 12.5 Differences in the anthropometric and performance-related characteristics of senior club (SC) forwards and backs. Level of significance was set at $P < 0.05$.

	Forwards (n = 16)	Backs (n = 9)
Height (cm) *	185.7(5.8)	178.0(7.6)
Body mass (kg) *	104.8(12.8)	84.0(7.3)
Body mass Index (kg.m^{-2}) *	30.4(3.5)	26.6(2.8)
10 m sprint (s)	1.78(0.07)	1.72(0.04)
Average 10 m Momentum (kg.m.s^{-1}) *	588.2(70.4)	488.0(40.0)
Aerobic Power (ml.kg.min^{-1})	47.1(6.5)	49.0(6.0)
Absolute 1RM bench press (kg)	111.7(19.3)	93.9(21.5)
Rel. 1RM bench press (Wilks Score)	66.8(10.6)	62.5(14.1)

* indicates a significant difference between forwards and backs

Table 12.6 Differences in the anthropometric and performance-related characteristics of junior state (JS) forwards and backs. Level of significance was set at $P < 0.05$.

	Forwards (n = 9)	Backs (n = 10)
Height (cm)	183.6(6.4)	175.9(5.7)
Body mass (kg)	90.7(10.0)	79.4(8.3)
Body mass Index (kg.m^{-2})	26.9(2.2)	25.7(2.0)
10 m sprint (s)	1.89(0.03)	1.86(0.04)
Average 10 m Momentum (kg.m.s^{-1})	480.3(53.1)	427.6(48.8)
Aerobic Power (ml.kg.min^{-1})	47.8(5.3)	49.1(6.6)
Absolute 1RM bench press (kg)	87.9(19.2)	83.7(10.4)
Rel. 1RM bench press (Wilks Score)	57.1(8.0)	56.8(9.5)

There were no significant differences between forwards and backs

4 DISCUSSION

The results of this study provided a comparison of descriptive, anthropometric measures and performance-related measures of three different standards of rugby union players and also looked at positional differences across these three playing levels. These results are important in order to understand the physical requirements of international representation and to identify weaknesses in sub-international players. Though findings that the body mass of the JS group was less than that of the SC and INT players are to be expected, it was interesting that aerobic power, despite a large age difference, was not significantly different between SC and JS. With regards to positional differences, results indicated that, across the SC and INT playing groups, forwards were significantly taller, heavier, stockier (as indicated by their greater BMI) and had greater momentum than backs. However, when investigating heavily muscled athletes like rugby players BMI can be problematic as it gives no indication of body composition (Jonnalagadda *et al.*, 2004).

4.1 Playing Level

The younger JS group was significantly lighter than the INT and SC groups. This finding was expected as the JS group was a much younger cohort than the other two groups. Although the body mass of the INT and SC groups was not significantly different, the INT group was about 5 kg heavier than the SC group. Large standard deviations may have been responsible for this finding not being statistically different. However, in a practical sense this 5 kg difference would most likely represent a noticeable difference on the field. It should also be noted that all positions in rugby were included in each group meaning that results from a front row prop were included with those of a winger. Consequently, some differences may have been diluted and not seen in this study. Though not measured, training age, the number of years involved in training, might also be responsible for the body mass difference between the younger JS group and the INT and SC groups. Weight training for rugby union involves considerable training with the purpose of increasing strength and muscle mass. The cumulative effects of this over the many more years it was expected that the INT and SC had been training, with comparison to the JS group, could be a contributor to body mass differences. Natural growth and development as the JS group grow from being adolescents into adulthood would also contribute to this difference. The reader should keep in mind throughout that this process of maturation may have contributed to differences in many of the measures found in this research. Though skinfolds measures were performed on the JS groups they were not available for the INT and SC groups. Such a measure would have helped provide more useful body composition information. Such research has been conducted on a group of elite professional rugby union players (Duthie *et al.*, 2006). This research also shows how an elite rugby union player's body composition changes throughout a playing season and provides a good description of the body composition of elite rugby union players.

Speed differences followed an expected pattern demonstrating that with an increase in playing level, sprinting speed, over 10 m, improved significantly. Further, the average momentum of the INT and SC groups was greater than the JS group. Perhaps the most surprising finding in this study was that the only significant difference in aerobic power was between the INT and SC groups. Though, like the body mass differences, changes that are not significantly different may still be of interest in a practical setting, the researchers were very interested to find that the INT group was not significantly aerobically fitter than the JS group. For strength and conditioning coaches, this finding should be worrying as it highlights a low aerobic capacity of the SC group that is lower than the JS group. In a recent review on the physiological profile of rugby union players, aerobic power was reported to be of the order 55 ml.kg^{-1}.min^{-1} (Duthie *et al.*, 2003). This finding is on a par with the INT group. However it suggests that those in the SC group are far below the standard required if they were attempting to play up to international standard. Perhaps, this 'senior slump' in aerobic capacity can be attributed to the JS groups being involved in structured training programs year round as they are involved in school sports together with many players from the JS group training for school, club, and state representative rugby teams. Further, in comparison to the INT group, it is likely that the SC group trained less frequently than the INT and SC groups, and may have spent limited training time on improving aerobic fitness. These suggestions are speculation as training programs were not monitored and not controlled for. Such an investigation into the proportion of training devoted to the different qualities of fitness may help to elucidate why such differences exist.

Finally, upper body strength also provided results that were both expected and then surprising. The reader is reminded that only the INT groups performed a 1RM test; the JS and SC groups performed a 3RM and their 1RM was estimated from this score (Baker, 2004; Nicholas and Baker, 1995). It was expected that the absolute upper body strength should incrementally improve as a function of playing level. It was also expected that a relative measure of strength (Wilks score) would show differences between the groups. Due again to physical maturity, training age, and experience, absolute upper body strength should be a significantly different variable between these three groups. Baker (2001, 2002) reports similar findings in different playing levels and ages of rugby league players. He considers this quality to be so important as to suggest that the younger less-experienced players focus on improving their upper body strength as measured by the 1RM bench press. In his study, the groups were not significantly different when compared on body mass, so it is unclear whether a relative upper body strength measure would have provided the same results. In the current research, a relative measure indicated the same differences seen in the absolute strength results.

4.2 Positional differences

Though the results showed that the two senior team forwards were stockier (as indicated by their greater BMI) than backs, the limitations of the BMI must be acknowledged when viewing this finding. The BMI does not provide any information regarding body composition and thus is not possible to determine whether the forwards' greater BMI reflected an increased accumulation of useful muscle mass or an increase in the amount of less functional fat mass (Jonnalagadda *et al.*, 2004). Regardless of these limitations, the results of this study are consistent with the literature (Quarrie *et al.*, 1995, 1996; Rigg and Reilly, 1988; Holmyard and Hazeldine, 1993) and support the view that being tall, heavy and able to produce large quantities of total body momentum are important determinants of success for forwards at all levels of rugby. Specifically, these characteristics appear beneficial for winning the ball during lineouts and scrums and for dominating close contact situations like rucks, mauls and tackles (James *et al.*, 2005; Deutsch *et al.*, 2002).

No significant differences between the forwards or backs were observed in any of the three groups for aerobic power and relative bench press strength between forwards and backs. The similarity in aerobic power and/or relative upper body strength appears consistent with a number of studies of INT level players (Holmyard and Hazeldine, 1993; Bell, 1980; Maud and Shultz, 1984; Ueno *et al.*, 1988). However, previous studies of SC level players have typically observed significantly greater aerobic power (Nicholas and Baker, 1995; Quarrie *et al.*, 1995, 1996) and greater relative upper body push muscular endurance (as assessed by the pushup) in backs than forwards.

Compared to the SC backs, the SC forwards had significantly greater absolute 1RM bench press strength. In contrast, no significant difference in 10 m sprint speed or absolute bench press strength was observed between forwards and backs in either of the groups. These results could reflect differences in training practices (Nicholas, 1997; Quarrie and Williams, 2002) and/or the specific requirements of elite versus local club level rugby (Deutsch *et al.*, 2002).

To date, little research has quantified the sprinting ability of rugby forwards and backs (Quarrie *et al.*, 1995, 1996; Rigg and Reilly, 1988). All three of these studies have assessed sprinting speed over distances of 30 m or more. The lack of assessment of short duration sprinting speed is surprising as the literature indicates that the mean duration of sprints in rugby is only 2–3 s (Docherty *et al.*, 1988; Duthie *et al.*, 2005), a time approximating sprints of 10–20 m. It may therefore be difficult to directly compare our results to the literature. Nevertheless, the results of the present study are consistent with the literature (Quarrie *et al.*, 1995, 1996; Rigg and Reilly, 1988), in that SC forwards were slower than SC backs, however this difference was only significant at the $p < 0.01$ level and not the criterion level of 0.05. In contrast, no significant positional-differences were observed for 10 m sprint speed in the INT or JS groups. The results of the current study are unable to answer the issue of whether backs would be significantly faster over longer (e.g., 30 m or 40 m) sprints, although as backs typically engage in longer sprints than forwards (Duthie *et al.*, 2005), such a hypothesis would appear plausible.

5 PRACTICAL APPLICATION

These results help in providing important information about physical and performance-related measures of different standards of rugby union players. They are consistent with the findings of much of the literature in that regardless of playing level, rugby forwards are typically significantly taller and heavier and have greater body mass index and total body momentum than backs. Such results can be useful, particularly for players below international level, by helping to provide training goals. They can also help for strength and conditioning coaches to design a training program for players of different levels. Though further investigation is required, this research provides interesting findings in regards to the SC group, suggesting those at that level striving to play at an international level should pay more attention to their aerobic power and relative upper body strength. Finally, these results suggest that elite rugby union may be becoming more like rugby league where the backs and forwards are quite homogenous in terms of their physiological abilities (Gabbett, 2002a,b) or alternatively, that more elite players perform greater volumes of training.

6 REFERENCES

Australian Coaching Council, 1988, Multistage Fitness Test: A Progressive Shuttle Run Test for the Prediction of Maximum Oxygen Uptake (Canberra: Australian Coaching Council).

Australian Institute of Sport, 2004, National protocols for the assessment of strength and power edited by Laboratory Standards Assistance Scheme (Canberra: Australian Institute of Sport), p. 29.

Baker, D., 2001, Comparison of upper-body strength and power between professional and college-aged rugby league players. *Journal of Strength and Conditioning Research*, **15**, 30–35.

Baker, D., 2002, Differences in strength and power among junior-high, senior-high, college-aged, and elite professional rugby league players. *Journal of Strength and Conditioning Research*, **16**, 581–585.

Baker, D., 2004, Predicting 1RM or sub-maximal strength levels from simple "Reps to Fatigue" (RTF) Tests. *Strength & Conditioning Coach*, **12**, 19–24.

Bell, W., 1980, Body composition and maximal aerobic power of rugby union forwards. *Journal of Sports Medicine and Physical Fitness*, **20**, 447–451.

Bishop, B. and Hume, P.A., 2000, *Guidelines for Athlete Assessment in New Zealand Sport* (Wellington: Sports Science New Zealand).

Deutsch, M.U., Kearney, G.A. and Rehrer, N.J., 2002, A comparison of competition work rates in elite club and "Super 12" Rugby. In *Science and Football IV: Proceedings of the Fourth World Congress of Science and Football*. edited by Spinks, W., Reilly, T. and Murphy, A. (London: Routledge), pp. 161–166.

Docherty, D., Wenger, H.A. and Neary, P., 1988, Time-motion analysis related to the physiological demands of rugby. *Journal of Human Movement Studies*, **14**, 269–277.

Duthie, G., Pyne, D. and Hooper, S., 2003, Applied physiology and game analysis of rugby union. *Sports Medicine*, **33**, 973–991.

Duthie, G., Pyne, D. and Hooper, S. 2005, Time motion analysis of 2001 and 2002 super 12 rugby. *Journal of Sports Sciences*, **23**, 523–530.

Duthie, G.M., Pyne, D.B., Hopkins, W.G., Livingstone, S. and Hooper, S.L., 2006, Anthropometry profiles of elite rugby players: Quantifying changes in Lean Mass. *British Journal of Sports Medicine*, **40**, 202–207.

Gabbett, T.J., 2002a, Influence of physiological characteristics on selection in a semi-professional first grade rugby league team: A case study. *Journal of Sports Sciences*, **20**, 399–405.

Gabbett, T.J., 2002b, Physiological characteristics of junior and senior rugby league players. *British Journal of Sports Medicine*, **36**, 334–339.

Grant, S., Corbett, K., Amjad, A.M., Wilson, J. and Aitchison, T., 1995, A comparison of methods of predicting maximum oxygen uptake. *British Journal of Sports Medicine*, **29**, 147–152.

Holmyard, D.J. and Hazeldine, R.J., 1993, Seasonal variations in the anthropometric and physiological characteristics of international rugby union players. In *Science and Football II: Proceedings of the Second World Congress of Science and Football* edited by Reilly, T., Clarys, J.P. and Stibbe, A. (London: E & FN Spon), pp. 21–26.

James, N., Mellalieu, S.D. and Jones, N.M.P., 2005, The development of position-specific performance indicators in professional rugby union. *Journal of Sports Sciences*, **23**, 63–72.

Jonnalagadda, S. S., Skinner, R. and Moore, L., 2004, Overweight athlete: Fact or fiction? *Current Sports Medicine Reports*, **3**, 198–205.

Leger, L. A. and Gadoury, C., 1989, Validity of the 20m shuttle run with 1 minute stages to predict VO₂max in adults. *Canadian Journal of Sport Science*, **14**, 21–26.

Leger, L. A., Mercier, D., Gadoury, C. and Lambert, J., 1988, The multistage 20 metre shuttle run test for aerobic fitness. *Journal of Sport Sciences*, **6**, 93–101.

Maud, P. J. and Shultz, B. B., 1984, The US National Rugby Team: A physiological and anthropometric assessment. *Physician and Sportsmedicine*, **12**, 86–94.

Mayes, R. and Nuttall, F. E., 1995, A comparison of the physiological characteristics of senior and under 21 elite rugby union players. *Journal of Sports Sciences*, **13**, 13–14.

Nicholas, C. W., 1997, Anthropometric and physiological characteristics of rugby union football players. *Sports Medicine*, **23**, 375–396.

Nicholas, C. W. and Baker, J. S., 1995, Anthropometric and physiological characteristics of first- and second-class rugby union players. *Journal of Sports Sciences*, **13**, p. 15.

Norton, K. and Olds, T., 1996, *Anthropometrica* (Marrackville: UNSW Press).

Quarrie, K. and Williams, S., 2002, Factors associated with pre-season fitness attributes of rugby players. In *Science and Football IV: Proceedings of the Fourth World Congress of Science and Football* edited by Spinks, W., Reilly, T. and Murphy, A. (London: Routledge), pp. 89–97.

Quarrie, K.L., Handcock, P., Toomey, M.J. and Waller, A.E., 1996, The New Zealand rugby injury and performance project. IV. anthropometric and physical performance comparisons between positional categories of senior a rugby players. *British Journal of Sports Medicine*, **30**, 53–56.

Quarrie, K.L., Handcock, P., Waller, A.E., Chalmers, D.J., Toomey, M.J. and Wilson, B.D., 1995, The New Zealand rugby injury and performance project. III. Anthropometric and physical performance characteristics of players. *British Journal of Sports Medicine*, **29**, 263–270.

Rigg, P. and Reilly, T., 1988, A fitness profile and anthropometric analysis of first and second class rugby union players. In *Science and Football: Proceedings of the First World Congress of Science and Football.* edited by Reilly, T. (Liverpool: E & FN Spon), pp. 194–200.

Sproule, J., Kunalan, C., Mcneill, M. and Wright, H., 1993, Validity of the 20-MST for predicting VO_2max of adult Singaporean athletes. *British Journal of Sports Medicine*, **27**, 202–204.

Ueno, Y., Watai, E. and Ishii, K., 1988, Aerobic and anaerobic power of rugby football players. In *Science and Football: Proceedings of the First World Congress of Science and Football* edited by Reilly, T. (Liverpool: E & FN Spon), pp. 201–205.

Vanderburgh, P.M. and Batterham, A.M., 1999, Validation of the Wilks powerlifting formula. *Medicine and Science in Sports and Exercise*, **31**, 1869–1875.

Young, W.B., Newton, R.U., Doyle, T.L.A., Chapman, D., Cormack, S., Stewart, G. and Dawson, B., 2005, Physiological and anthropometric characteristics of starters and non-starters and playing positions in elite Australian rules football: A case study. *Journal of Science and Medicine in Sport*, **8**, 333–345.

CHAPTER THIRTEEN

Direct-depth measurement of subcutaneous adipose tissue

M.J. Marfell-Jones[1], S. Provyn[2] and J.P. Clarys[2]

[1]Universal College of Learning, Palmerston North, New Zealand.
[2]Experimental Anatomy, Vrije Universiteit Brussel, Belgium.

1 INTRODUCTION

Unlike Computed Tomography (CT) or Magnetic Resonance Imaging (MRI) scanning, skinfold measurement to predict body composition is widespread, predominantly because of its affordability and portability. The validity of skinfold measurement as an accurate predictor of whole-body adiposity, however, has long been questioned (e.g. Flint *et al.*, 1977; Hägar, 1981; Martin *et al.*, 1985, 1994; Clarys *et al.*, 1987; Marfell-Jones *et al.*, 2003). Many investigators have demonstrated the difference in values for subcutaneous and/or whole-body adiposity predicted from skinfold measurement and those values predicted from other methods on the same subjects (e.g. Weits *et al.*, 1986; Kuczmarski *et al.*, 1987; Hayes *et al.*, 1988; Stevens-Simon *et al.*, 2001; Pontiroli *et al.*, 2002; Snijder *et al.*, 2002; Tothill and Stewart, 2002) resulting in reasonable unanimity that skinfold measurements are not consistently reliable predictors of whole-body adiposity, particularly in the large to obese subject.

To consider this issue, and as a precursor to investigating a more-reliable method for predicting whole-body adiposity, data from the Brussels Cadaver Analysis Study (CAS) were analysed to confirm the strength of the relationship between subcutaneous and whole-body adipose tissue, and to compare the measurement of subcutaneous adipose tissue by skinfold measurement and by direct-depth measurement. The extent to which each of these two techniques correlated with internal adiposity and whole-body adiposity was also examined.

2 METHODS

The Brussels Cadaver Analysis Study (CAS) consisted of a three-part cadaver dissection study undertaken at the Vrije Universiteit Brussel (VUB) between 1979 and 1991. Details of these whole-body dissection projects have been reviewed by Clarys *et al.* (1999). For the purposes of this investigation, the majority of data came from the first part of the study (Clarys *et al.*, 1984; Martin, 1984) with

180   *Marfell-Jones* et al.

additional subject data being added from the second part (Marfell-Jones, 1984; Clarys and Marfell-Jones, 1986) and the third part (Janssens *et al.*, 1994), where possible. In the first part, extensive anthropometry was carried out bilaterally on each of 25 cadavers. This included the measurement of eight skinfolds according to the protocol of Martin and Saller (1957), which protocol was subsequently adopted by the International Society for the Advancement of Kinanthropometry (2001). Following the skinfold measurements, a short incision was made, in the direction of the skinfold, at each of the skinfold sites, to the depth of the muscle fascia. The distance from the surface of the skin to the muscle fascia was then measured by inserting a small metal ruler into the incision to the depth of the fascia (with minimal pressure) and reading the depth measurement with the two sides of the incision closed back in juxtaposition against the ruler. Each cadaver was then segmented into head and neck, trunk, left and right upper limbs, and left and right lower limbs. Following segmentation, each segment was further dissected into skin, adipose tissue, skeletal muscle, bone, and organ and viscera masses. The weight of any fluid separating from the tissue was added back to the tissue weight. All tissues were stored in airtight, humidified containers until weighing. The evaporative weight loss occurring through the dissection process (taken to be the difference between pre-dissection body weight and the sum of all tissues after the dissection) was added back to each component in proportion to its weight. A complete dissection lasted from 10–15 hours and required a team of at least 12 people. The whole body was weighed in both air and water, as were the segments and the masses. These actions generated both skinfold and direct-depth measurements at each of eight sites, plus values for the amounts of adipose tissue located subcutaneously and internally (which summed to the total amount of adipose tissue for the whole body) in each cadaver. Internal adipose tissue included dissectable intramuscular adipose tissue as well as dissectable adipose tissue from within the trunk. The relationships between these several entities were then examined.

3 RESULTS AND DISCUSSION

Descriptive characteristics for the main sample are shown in Table 13.1. The anthropometric characteristics of the sample prior to dissection did not differ significantly from living Belgians of similar age (Marfell-Jones *et al.*, 2003).

Means, standard deviations and ranges for the three adipose tissue masses, the sums of eight direct depths and eight skinfolds respectively, and the subcutaneous and internal AT masses expressed as a percentage of total AT mass are shown in Table 13.2.

Table 13.1 Descriptive characteristics of 12 male and 13 female cadavers. [Reproduced from Martin *et al.*, 2003 with permission.]

Variable	Group	Mean	s.d.	Range
Age (yr)	men	71.7	8.5	55.0–83.0
	women	79.7	7.4	68.0–94.0
	combined	75.8	8.7	55.0–94.0
Stature (cm)*	men	168.9	8.0	159.2–186.5
	women	159.5	6.4	151.3–172.3
	combined	163.7	8.5	151.3–186.5
Weight (kg)	men	66.2	12.5	51.7–88.9
	women	62.5	9.4	48.2–75.4
	combined	64.3	10.9	48.2–88.9
% Adipose Tissue	men	28.1	6.4	17.8–43.9
	women	40.5	7.4	28.6–54.1
	combined	34.6	9.3	17.8–54.1

*estimated from supine length

Table 13.2 Adipose tissue masses, skinfold sums and direct depth sums, and percentages of adipose tissue located subcutaneously and internally in 11 male and 12 female cadavers.

Variable	Group	Mean	s.d.	Range
Subcutaneous AT mass (kg)	men	11.1	3.4	5.5–17.0
	women	17.9	5.4	9.5–27.9
	combined	14.6	5.6	5.5–27.9
Internal AT mass (kg)	men	7.5	1.9	4.2–9.9
	women	8.5	2.9	4.9–13.6
	combined	8.0	2.5	4.2–13.6
Total AT mass (kg)	men	18.6	4.9	9.7–25.7
	women	26.4	7.8	14.4–40.1
	combined	22.7	7.6	9.7–40.1
Sum 8 skinfolds left (mm)	men	100.6	29.7	55.7–142.3
	women	156.3	38.5	88.2–212.0
	combined	127.0	43.8	55.7–212.0
Sum 8 direct depths (skin + AT) (mm)	men	101.0	32.0	66.0–163.5
	women	167.1	45.6	106.5–246.5
	combined	135.5	51.4	66.0–246.5
% AT located subcutaneously	men	59.4	6.0	46.1–66.3
	women	67.8	4.3	57.5–73.4
	combined	63.8	6.6	46.1–73.4
% AT located internally	men	40.6	5.8	33.7–53.9
	women	32.2	4.3	26.6–42.5
	combined	36.2	6.6	26.6–53.9

Though the preponderance of adiposity lay subcutaneously in both sexes, the women carried a significantly greater percentage of their adipose tissue subcutaneously than did the men (and, concomitantly, a significantly lesser percentage internally).

Correlational analysis revealed good to almost-perfect relationships between the three adipose tissue masses in both men and women (see Table 13.3).

Table 13.3 Correlations between external, internal and whole-body adipose tissue masses in men and women.

	Men ($n = 15$)	Women ($n = 17$)
Subcutaneous AT vs Internal AT	0.72	0.86
Subcutaneous AT vs Whole-body AT	0.96	0.97
Internal AT vs Whole-body AT	0.85	0.89

It should not be overlooked, however, that in the correlations of subcutaneous and internal AT masses with whole-body AT mass, the former, in each case, is a subset of the latter, and that when a subset constitutes by far the majority of a superset (as is the case with subcutaneous adipose tissue mass) a high correlation is inevitable. Nevertheless, in our view, the results clearly indicate that both internal and whole body adipose tissue masses can be confidently predicted from subcutaneous mass, irrespective of gender, and the challenge becomes one of accurately measuring or predicting the subcutaneous adipose tissue mass.

We compared two subcutaneous adipose tissue mass predictors – skinfold measures and direct depth measures. The relationships identified between the former, i.e. the skinfold measures (expressed as a sum of skinfolds) and subcutaneous, internal and whole-body adipose tissue masses, are shown in Table 13.4.

Table 13.4 Correlations between sum of skinfolds and subcutaneous, internal and whole-body adipose tissue masses and direct depth measures.

	Men ($n = 11$)	Women ($n = 12$)
Sum skinfolds vs Subcutaneous AT	0.82	0.56
Sum skinfolds vs Internal AT	0.69	0.27
Sum skinfolds vs Whole-body AT	0.86	0.49
Sum skinfolds vs Direct depths	0.82	0.54

Inspection of the results immediately revealed gender differences in the strength of the relationships. Whereas the correlations between entities were good in men, they were significantly weaker in women, leading to greatly-reduced confidence in our ability to predict subcutaneous, internal or whole body fatness in females from skinfold measures. [The relationship of particular interest, skinfold measurements and subcutaneous fat, is shown in Figure 13.1.] The observed differences between the two groups was, in our view, due to skinfold compressibility.

The issue of skinfold compressibility is well established in the literature (e.g. Martin, 1984; Jones *et al.*, 1979; Beceque *et al.*, 1986; Clarys *et al.*, 1987; Martin *et al.*, 1992; Ward and Anderson, 1993; Marfell-Jones *et al.*, 2003). In our sample, it was postulated that differences in skinfold compressibility between the two sexes might account for the lower correlations in the females. However, analysis of compressibility across all skinfold sites in the CAS sample showed that minor differences identified between the two sexes were not significant, a point already made by Martin (1984) and Clarys *et al.* (1987).

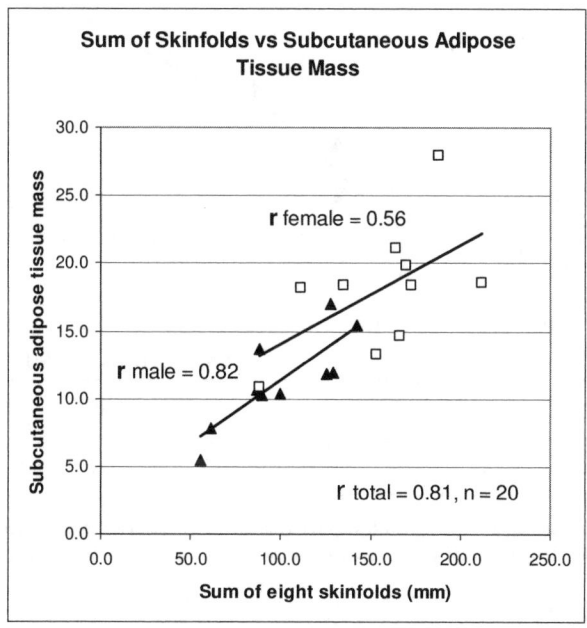

Figure 13.1 Correlations between sum of 8 skinfolds & subcutaneous adipose tissue mass in males compared to females.

Significant differences in compressibility were identified, however, in relation to the total amount of dissectible adipose tissue. Correlation of the sum of skinfolds with the total dissectible adipose tissue mass of the ten subjects with the largest adipose tissue mass (which included two men) and of the ten subjects with the smallest adipose tissue mass (which included two women) revealed correlations of 0.39 and 0.81 respectively (Figure 13.2), indicating that compressibility variations may well be related to adipose tissue amount rather than sex.

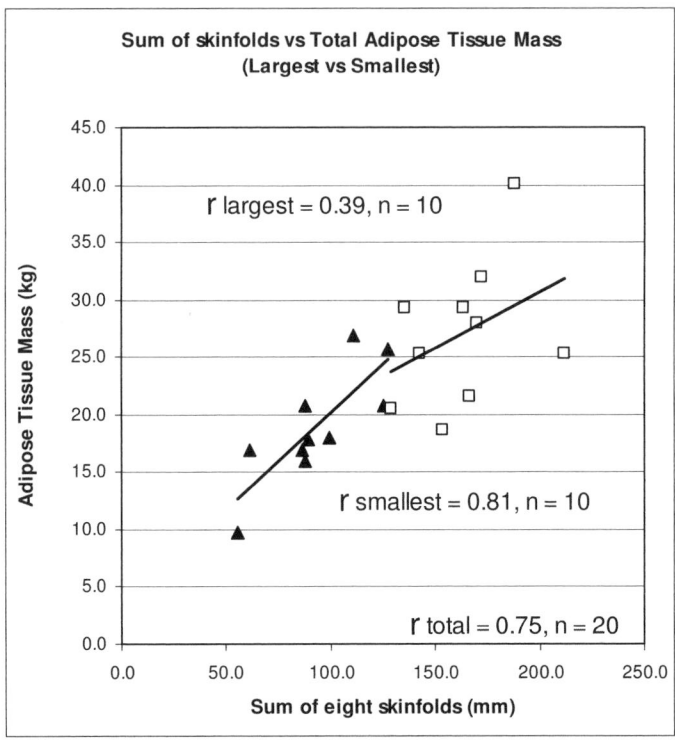

Figure 13.2 Correlations between sum of 8 skinfolds and total adipose tissue mass in ten subjects with greatest amount of total adipose tissue compared to ten subjects with least amount.

This finding reinforces that of Martin (1984), on a lesser number of cadavers, and is consistent with our own observational experience over many years of taking skinfold measurements. [The fact that by far the majority of women, of course, have greater amounts of adipose tissue than do the majority of men, obviously has a tendency to obscure this relationship.] Further, it is our view, based on experience, that compressibility is also related to the relative state of "fatness" of the individual. The skinfolds of subjects who have substantially reduced their adiposity below its previous maximum, exhibit greater skinfold compressibility than do those of subjects whose adiposity is at, or near, its maximum to date. We believe this is due to a greater mobility of the subcutaneous adipose tissue when compressed, which can be experienced by finger and thumb touch as well as observation of a still-moving calliper needle.

The relationships between direct depth (skin plus adipose tissue) measures and subcutaneous, internal and whole-body adipose tissue masses, are shown in Table 13.5. These demonstrate strong correlations in both men and women.

Table 13.5. Correlations between direct depth measures, subcutaneous, internal, and whole-body adipose tissue in men and women.

	Men ($n = 11$)	Women ($n = 12$)
Σ8 Direct depths vs Subcutaneous AT	0.83	0.92
Σ8 Direct depths vs Internal AT	0.73	0.68
Σ8 Direct depths vs Whole-body AT	0.87	0.88

The relationship of particular interest, the sum of eight direct depth (skin plus adipose tissue) measures and subcutaneous adipose tissue, is shown in Figure 13.3.

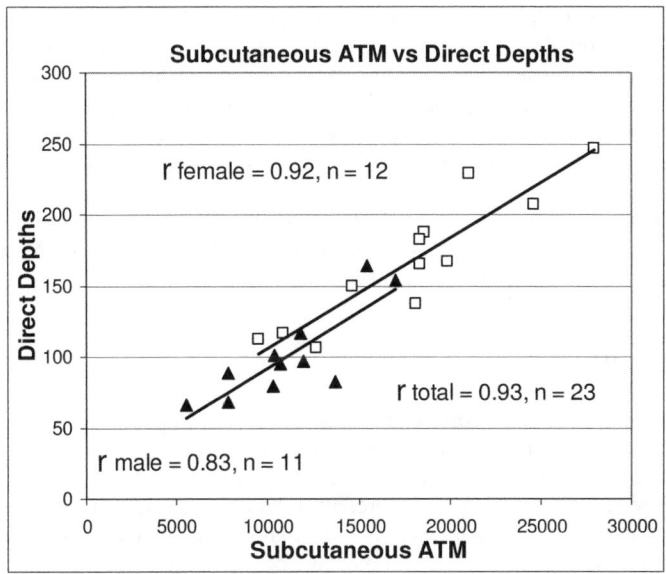

Figure 13.3 Correlations between direct depths and subcutaneous adipose tissue mass in males compared to females.

These findings are extremely encouraging, as they confirm that, unlike skinfold measurements, the measurement of direct-depth will allow the prediction of both subcutaneous and whole-body adipose tissue masses with confidence in both sexes.

Fortunately, neither direct measurement of subcutaneous adipose tissue mass nor direct measurement of depth of skin plus adipose tissue, as conducted in our cadaver research, is an option in vivo, so we must look to other techniques as possibilities for general measurement of one or both of these entities. Amongst the techniques currently espoused for such measurements, X-ray, Computed Tomography (CT), Dual Energy X-Ray Absorptiometry (DEXA or DXA),

Magnetic Resonance Imaging (MRI) and Ultrasound (US) are among the best known.

It was not the purpose of this paper to conduct an in-depth review of these techniques, but some general comments can be made. The first two (X-ray and CT) can be discounted as viable options because of their invasive nature. [Of course they are used extensively on the injured or ill, but only when the negative aspects of their use are considered a lesser priority that the patient's welfare.] Of the remaining three, the accuracy of DEXA to measure subcutaneous adipose tissue depth has been endorsed by a number of authors (e.g. Taaffe *et al.*, 1994; Snijder *et al.*, 2002; Ball *et al.*, 2004), but questioned by others (e.g. Clasey *et al.*, 1999; Black *et al.*, 2002; Van Der Ploeg *et al.*, 2003). The use of MRI is supported by some (e.g. Fowler *et al.*, 1991; Sobol *et al.*, 1991), yet questioned by others (e.g. Hayes *et al.*, 1998). The use of US is supported by some (e.g. Balta *et al.*, 1981; Jones *et al.*, 1986; Bellisari *et al.*, 1993; Yasukawa, 1995), yet questioned by others (e.g. Ramirez, 1992; Orphanidou *et al.*, 1994; Hayes *et al.*, 1998; Stevens-Simon *et al.*, 2001). However, except for Balta *et al.* (1981), who made direct measurements on patients immediately preceding laparotomy surgery, and Jones *et al.* (1986), who made direct measurements on a single cadaver, in none of the above papers is a "gold standard" reference, such as cadaver research, used to verify the assertions made therein. Therefore, in our view, more conclusive evidence is needed as to the accuracy of any or all of these three techniques before we can advocate their use to measure subcutaneous adipose tissue depth (and hence predict subcutaneous adipose tissue mass).

4 CONCLUSION

We conclude from our findings that subcutaneous adipose tissue mass is an excellent predictor of total body adipose tissue mass, but that, of the two predictors of subcutaneous adipose tissue mass evaluated, the first, skinfold measurement, cannot be relied on to adequately predict subcutaneous adipose tissue mass in "fatter" subjects, which group includes the majority of women. [We emphasise, however, that whereas we do not advocate the use of skinfold measurement to predict total body adipose tissue, we do not discount skinfold measurement as a valuable tool for intra-individual comparison of adiposity. Concomitantly, we reiterate our earlier advice (Clarys *et al.*, 1987; Marfell-Jones *et al.*, 2003), that such measures should stand alone and not be transformed into predictions of % body fat, or whole-body adiposity.] Direct depth measurement, on the other hand, has been shown to predict subcutaneous adipose tissue mass in both sexes, irrespective of the level of "fatness". The major reasons for the continued use of skinfold measures in the face of on-going evidence of their limitations, are the cost and portability of the equipment used. Currently, DEXA, MRI and US equipment is expensive and only the last is relatively portable. We therefore add these elements to our search criteria and look forward to the development of an accurate, economic, portable piece of equipment that can subsume the ubiquitous role that skinfold measurement currently occupies.

5 REFERENCES

Ball, S.D., Altena, T.S. and Swan, P.D., 2004, Comparison of anthropometry to DXA: a new prediction equation for men. *European Journal of Clinical Nutrition*, pp. 1525–1531.

Balta, P.J., Ward, M.W.M. and Tomkins, A.M., 1981, Ultrasound for measurement of subcutaneous fat. *The Lancet*, Feb 28, pp. 504–505.

Beceque, M.D., Katch, V.L. and Moffatt, R.J., 1986, Time course of skin-plus-fat compression in males and females. *Human Biology*, **58**(1), 33–42.

Bellisari, A., Roche, A.F. and Siervogel, R.M., 1993, *Reliability* of *B-mode* ultrasonic measurements of subcutaneous adipose tissue and intra-abdominal depth: comparisons with skinfold thicknesses. *International Journal of Obesity and Related Metabolic Disorders,* Aug, **17**(8), 475–480.

Black, E., Petersen, L., Kreutzer, M., Toubro, S., Sorensen, T.I., Pedersen, O. and Astrup, A., 2002, Fat mass measured by DXA varies with scan velocity. *Obesity Research*,**10** (2), 69–77.

Clarys, J.P. and Marfell-Jones, M.J., 1986, Anthropometric prediction of component tissue masses in the minor limb segments of the human body. *Human Biology*, **58**(5), 761–769.

Clarys, J.P., Martin, A.D. and Drinkwater, D.T., 1984, Gross tissue weights in the human-body by cadaver dissection. *Human Biology*, **56**(3), 459–473.

Clarys, J.P., Martin, A.D. and Drinkwater, D.T., 1988, Physical and structural distribution of human skin. *Human Biology*, Budapest, **18**, 55–63.

Clarys, J.P., Martin, A.D., Drinkwater, D.T., and Marfell-Jones, M.J., 1987, The skinfold: myth and reality. *Journal of Sports Science*, **5** (1), 3–33.

Clarys, J.P., Martin, A.D., Marfell-Jones, M.J, Janssens, V., Caboor, D. and Drinkwater, D.T., 1999, Human body composition: A review of adult dissection data. *American Journal of Human Biology*, **11**, 167–174.

Clasey, J.L., Kanaley, Wideman, L., Heymsfield, S.B., Teates, C.D., Gutgesell, M.E., Thorner, M.O., Hartman, M.L. and Weltman, A., 1999, Validity of methods of body composition assessment in young and older men and women. *Journal of Applied Physiology*, **86**(5), 1728–1738.

Flint, M.M., Drinkwater, B.L., Wells, C.L. and Horvath, S.M., 1977, Validity of estimating body fat of females: Effect of age and fitness. *Human Biology*, **49**(4), 559–572.

Fowler, P.A., Fuller, M.F., Glasbey, C.A., Foster, M.A., Cameron, C.G., McNeill, G. and Maughan, R.J., 1991, Total and subcutaneous adipose tissue in women: the measurement of distribution and accurate prediction of quantity by using magnetic resonance imaging. *American Journal of Clinical Nutrition*, **54**(1), 18–25.

Hayes, P.A., Sowood, P.J., Belyavin, A., Cohen, J.B. and Smith, F.W., 1988, Sub-cutaneous fat thickness measured by magnetic resonance imaging, ultrasound, and calipers. *Medicine and Science in Sports and Exercise*, Jun, **20**(3), 303–309.

Hägar, A., 1981, Estimation of body fat in infants, children and adolescents. In: Bonnet, P. (Ed.) *Adipose Tissue in Childhood*, pp. 49–56 CRC Press, Boca Raton, Fl. USA.

International Society for the Advancement of Kinanthropometry (ISAK), 2001, *International Standards for Anthropometric Assessment* (Underdale, SA, Australia).

Janssens, V., Thys, P., Clarys, J.P., Kvis, H., Chowdhury, B., Zinzen, E. and Cabri, J., 1994, Post-mortem limitations of body composition analysis by computed tomography. *Ergonomics*, **37**, 207–216.

Jones, P.R., Davies, P.S. and Norgan, N.G., 1986, Ultrasonic measurements of subcutaneous adipose tissue thickness in man. *American Journal of Physical Anthropology*, **71**(3), 359–363.

Jones, P.R.M., Marshall, W.A. and Harding, R.H., 1979, Harpenden electronic readout (Hero) caliper and its application in the study of skinfold compressibility. *Annals of Human Biology*, **6**(3), 291.

Kuczmarski, R.J., Fanelli, M.T. and Koch, G.G., 1987, Ultrasonic assessment of body composition in obese adults: Overcoming the limitations of the skinfold caliper. *American Journal of Clinical Nutrition*, Apr, **45**(4), 717–724.

Marfell-Jones, M.J. 1984, An anatomically-validated method for the anthropometric prediction of segmental masses. *Doctoral Thesis*, Simon Fraser University.

Marfell-Jones, M.J., Clarys, J.P., Alewaeters, K., Martin, A.D. and Drinkwater, D.T., 2003, Effects of skin thickness and skinfold compressibility on skinfold thickness measurement. *Biometrie Humaine et Anthropologie*, **21**(1–2), 103–117.

Martin, A.D., 1984, An anatomical basis for assessing human body composition: Evidence from 25 dissections. PhD Thesis, Simon Fraser University, Burnaby, Canada (and Vrije Universiteit Brussel).

Martin, A.D., Daniel, M., Drinkwater, D.T. and Clarys, J.P., 1994, Adipose-tissue density, estimated adipose lipid fraction and whole-body adiposity in male cadavers. *International Journal of Obesity*, **18**(2), 79–83.

Martin, A.D., Drinkwater, D.T., Clarys, J.P., Daniel, M. and Ross, W.D., 1992, Effects of skin thickness and skinfold compressibility on skinfold thickness measurement. *American Journal of Human Biology*, **4**(4), 453–460.

Martin, A.D., Janssens, V., Caboor, D., Clarys, J.P. and Marfell-Jones, M.J., 2003, Relationship between visceral, trunk, and whole-body adipose tissue weights by cadaver dissection. *Annals of Human Biology*, **30**(6), 668–677.

Martin, A.D., Ross, W.D., Drinkwater, D.T. and Clarys, J.P., 1985, Prediction of body-fat by skinfold caliper – Assumptions and cadaver evidence. *International Journal of Obesity*, **9**, 31–39 Suppl.1.

Martin, R. and Saller, K., 1957, Lehrbuch der Anthropologie. R. Gustav Fischer, Stuttgart.

Orphanidou, C., McCargar, L., Birmingham, C.L., Mathieson, J. and Goldner, E., 1994, Accuracy of subcutaneous fat measurement: Comparison of skinfold callipers, ultrasound, and computed tomography. *Journal of the American Dietetic Association*, **94**(8), 855–858.

Pontiroli, A.E., Pizzocri, P., Giacomelli, M., Marchi, M., Vendani, P., Cucchi, E., Orena, C., Folli, F., Paganelli, M. and Ferla, G., 2002, Ultrasound measurement of visceral and subcutaneous fat in morbidly obese patients before and after laparoscopic adjustable gastric banding: Comparison with computerized tomography and with anthropometric measurements. *Obesity Surgery*, Oct, **12**(5), 648–651.

Ramirez, M.E., 1992, Measurement of subcutaneous adipose tissue using ultrasound images. *American Journal of Physical Anthropology*, Nov, **89**(3), 347–357.

Snijder, M.B., Visser, M., Dekker, J.M., Seidell, J.C., Fuerst, T., Tylavsky, F., Cauley, J., Lang, T., Nevitt, M. and Harris, TB., 2002, The prediction of visceral fat by dual-energy X-ray absorptiometry in the elderly: A comparison with computed tomography and anthropometry. *International Journal of Obesity*, Jul, **26**(7), 984–993.

Sobol, W., Rossner, S., Hinson, B., Hiltbrandt, E., Karstaedt, N., Santago, P., Wolfman, N., Hagaman, A. and Crouse 3rd, J.R., 1991, Evaluation of a new magnetic resonance imaging method for quantitating adipose tissue areas. *International Journal of Obesity*, **15** (9), 589–599.

Stevens-Simon, C., Thuren, P., Barett, J., and Stamm, E., 2001, Skinfold caliper and ultrasound assessments of change in the distribution of subcutaneous fat during adolecent pregnancy. *International Journal of Obesity*, Sep, **25**(9), 1340–1346.

Taaffe, D.R., Lewis, B. and Marcus, R. 1994. Regional fat distribution by dual-energy X-ray absorptiometry: Comparison with anthropometry and application in a clinical trial of growth hormones and exercise. *Clinical Science*, **87**(5), 581–586.

Tothill, P. and Stewart, A.D., 2002, Estimation of thigh muscle and adipose tissue volume using magnetic resonance imaging and anthropometry. *Journal of Sports Science*, Jul, **20**(7), 563–576.

Van der Ploeg, G.E., Withers, R.T. and Laforgia, J. 2003. Percent body fat via DEXA: Comparison with a four-compartment model. *Journal of Applied Physiology*, **94** (2), 499–506.

Ward, W., and Anderson, G., 1993, Examination of the skinfold compressibility and skinfold thickness relationship. *American Journal of Human Biology*, **5**(5), 541–548.

Weits, T., van der Beek, E.J. and Wedel, M., 1986, Comparison of ultrasound and skinfold caliper measurement of subcutaneous fat tissue. *Internatonal Journal of Obesity*, **10**(3), 161–168.

Yasukawa, M., Horvath, S.M., Oishi, K., Kimura, M., Williams, R. and Maeshima, T. 1995. Total body fat estimations by near infra-red interactance, A-mode ultrasound, and underwater weighing. *Applied Human Sciences*, **14**(4), 183–189.

CHAPTER FOURTEEN

The role of physical activity in the prevention and treatment of obesity as an inflammatory condition: Review article

J. Beneke[1], C. Underhay[1], A.E. Schutte[2], and J.H. De Ridder[1]

[1]School for Biokinetics, Recreation and Sport Science, North-West University (Potchefstroom Campus), Potchefstroom, South Africa

[2] School for Nutrition, Physiology and Consumer Sciences, North-West University (Potchefstroom Campus), Potchefstroom, South Africa

1 INTRODUCTION

Cross-sectional epidemiological studies and interventions have demonstrated the benefit of physical activity in the primary prevention of coronary heart disease, diabetes mellitus and hypertension (Grundy et al., 2004; Sobngwi et al., 2002; Grundy et al., 1999; Leon, 1997). Improved functional capacity, flexibility, muscle strength and endurance are some of the main measurable outcomes of physical activity. Bouchard and Blair (1999) reported that regular physical activity plays a vital role in the regulation of energy balance. Physical activity also reduces the risk of being affected by the co-morbidities of obesity and results in lower all-cause and cardiovascular death rates (Bouchard and Blair, 1999).

Obesity is considered one of the cornerstone risk factors that cluster together when describing the metabolic syndrome (MS). According to Lam et al. (2003), most of the variables in the metabolic syndrome result from multiple factors linked by adiposity. These variables include hypertriglyceridemia, low levels of high-density lipoprotein cholesterol (HDL-C), hypertension, dysfibrinolysis, inflammation (associated with elevated C-reactive protein [CRP]), and/or elevated fasting insulin (Appel et al., 2004).

Obesity-related inflammatory markers (CRP, Interleukin-6 [IL-6] and tumor necrosis factor-α [TNF-α]) may be important mediators in the pathophysiology of coronary heart disease (CHD) and Type 2 diabetes (Pischon et al., 2003; Rexrode et al., 2003). Significant inverse relationships between physical activity and these inflammatory markers (IL-6, TNF-α and CRP) have been identified by researchers, who also suggested that the beneficial association could partially be due to less body fat (Aronson *et al.*, 2004a; Pischon *et al.*, 2003; Church *et al.*, 2002; La Monte *et al.*, 2002; Wannamethee *et al.*, 2002).

Although many studies have already been undertaken on physical activity and obesity, the purpose of this review was to evaluate existing literature to determine if physical activity played a major role in the treatment and prevention of inflammation associated with obesity and related to the metabolic syndrome.

2 METHODS

For this review, a computer-assisted literature search was utilized to identify research between 1995 and 2005. The following databases were used: NEXUS, Science Direct, PubMed and Medline. Search keywords related to physical activity (fitness, exercise and training); obesity (overweight, body mass index, waist circumference and waist-to-hip ratio) and inflammatory markers (CRP, IL-6, chronic inflammation, low-grade inflammation). References identified by previous reviewers (not identified as part of the computerized literature search) were also considered.

Since a general consensus already existed that obesity was part of the metabolic syndrome this review article limited itself to studies which focused on obesity as a state of chronic low-grade inflammation. A strong emphasis was placed on the relationship between physical activity and CRP and IL-6. A total of 39 cross-sectional and clinical-trial studies were identified that met the above inclusion criteria. Of these, 13 studies demonstrated a relationship between obesity and inflammatory markers (Table 14.1), 15 studies identified the effects of decreased obesity by weight loss on systemic markers of inflammation (Table 14.2) and 16 studies showed an association between markers of inflammation and physical activity or fitness (Table 14.3). Studies listed include author, group and year of publication, number of participants with mean age, BMI and other measures of obesity and the effect on inflammatory markers.

Table 14.1 Summary of published data on the relationship between obesity and inflammatory markers.

Author group and year of publication	Participants (n) and age (yr)	Mean BMI (kg/m²), WC (cm), and WHR	Effects on inflammatory markers by measure of obesity
Research on CRP:			
Heilbronn et al., 2001	83 healthy obese women Age 48.0 ± 0.9	BMI 33.8 ± 0.4 WC 98.3 ± 1.0 WHR 0.83 ± 0.01	↑CRP ↑BMI ↑WC
Lemieux et al., 2001	159 adult men Age 22 to 63	BMI 30.3 ± 3.9 WC 101.0 ± 9.3	↑CRP ↑BMI ↑WC ↑visceral AT
Hak et al., 1999	186 healthy women Age 50.9 ± 2.3	BMI 24.9 ± 4.0 WC 81.5 ± 9.5 WHR 0.77 ± 0.05	↑CRP ↑BMI ↑WC
Aronson et al., 2004a	892 subjects Age 50 ± 10 63% Males	BMI 28.2 ± 4.9	↑CRP ↑BMI
Vikram et al., 2003	332 male * Age 18.2 ± 2.3 46 female * Age 16.9 ± 2.0	BMI 20.1 ± 3.3 BMI 19.9 ± 3.3	↑CRP ↑BMI ↑Fat% ↑WHR
Weyer et al., 2002	32 males and females $ Age 18 and 43	BMI 20.2 ± 55.8	↑CRP ↑BMI ↑Fat%
Chambers et al., 2001	507 # Subjects Age 49.4 ± 6.5 518 * Subjects Age 49.0 ± 6.9	BMI 26.7 ± 4.0# WHR 0.93 ± 0.07# BMI 26.9 ± 3.5 * WHR 0.97 ± 0.07 *	↑CRP in * ↑CRP ↑BMI ↑WHR
Rutter et al., 2004	1681 women 1356 men Mean age 54	BMI 26 ± 5 women WC 34 ± 5 BMI 28 ± 4 men WC 39 ± 4	↑CRP with increased features of the metabolic syndrome
Barinas-Mitchell et al., 2001 Healthy Women Study	101 HRT Age 59.5 ± 2.1 106 HRTN Age 59.3 ± 1.8	BMI 26.8 ± 4.5 (HRT) WHR 0.79 ± 0.07 BMI 28.0 ± 6.1 (HRTN) WHR 0.79 ± 0.09	↑CRP ↑BMI ↑WC ↑WHR ↑Fat%
Tchernof et al., 2002	61 obese women Age 56.4 ± 5.2	BMI 35.6 ± 5.0	↑CRP ↑BMI ↑WHR
Research on IL-6:			
Vozarova et al., 2001	58 men and women without diabetes $ Age 30 ± 7	BMI 32.5 ± 6.5 WC 104.14 ± 15.24	↑IL-6 ↑Fat% ↑WC
Ziccardi et al., 2002	56 obese women Age 35.3 ± 4.8	BMI 37.2 ± 2.2 WHR 0.84 ± 0.06	↑IL-6 ↑WHR

Note: ↑ = increased, BMI = mean body mass index WHR = waist-hip-ratio WC = waist circumference ± standard error of the mean, CRP = C-reactive protein, IL = interleukin, Visceral AT = Visceral adipose tissue, HRT = Hormone replacement therapy users, HRTN = Hormone replacement therapy non users, # Caucasian, * Asian Indians, $ Pima Indians.

Table 14.2a Summary of published data on the effects of decreased obesity (by weight loss and/or physical activity) on systemic markers of inflammation.

Author group and year of publication	Participants and mean BMI, kg/m²	Intervention for decreasing obesity	Dura-tion	Magnitude of weight loss	Effects on inflammatory markers
Randomized, controlled trials:					
Intervention group (versus control group)					
Esposito et al., 2003	120 obese women BMI 35.0 ± 2.3 (v. BMI 34.7 ± 2.4)	1300–1500 kcal/d energy-restricted Mediterranean-style American Heart Association Step 1 diet (v. normal diet)	2 yr	14.7% of weight (v. 3.2% of weight)	CRP ↓ 34%; IL-6 ↓ 33% (v. no changes in CRP, IL-6)
Nicklas et al., 2004	316 older adults BMI 34.5 ± 5.4 (v. BMI 34.2 ± 5.0)	Behavioural counselling to achieve and keep a 5% weight loss (v. normal diet)	18 mo	5.1% of weight (v. 1.8% of weight)	CRP ↓ 3%; IL-6 ↓ 11%; sTNFR1 ↓ 2%; no change in TNF-α (v. no changes)
Intense, short-term dietary restrictions					
Xydakis et al., 2004	40 obese adults with metabolic syndrome BMI 38.9 ± 1.0	600–800 kcal/d very-low-calorie diet	4-6 wk	7.0% of weight	CRP↓14% No change in TNF-α
Heilbronn et al., 2001	83 obese women BMI 33.8 ± 0.4	Low-fat (15%), 1260 kcal/d	12 wk	7.9kg	CRP↓26%
Gallistl et al., 2001	49 obese children BMI 26.7 ± 1.4	908–1194 kcal/d energy-restricted diet	3 wk	5.2% of BMI, 3.1% of fat mass	IL-6 ↓ 49%
Bastard et al., 2000a	14 obese women BMI 39.5 ± 1.1	941 kcal/d very-low-calorie diet	3 wk	5.3% of BMI 8.5% of fat mass	IL-6 ↓ 17% no change in CRP, TNF-α
Bastard et al., 2000b	17 obese women BMI 39.9 ± 1.6	941 kcal/d very-low-calorie diet	3 wk	5.0% of fat mass	sTNFR1 ↓ 8% no change in sTNFR2

Note: ↓ = decreased, BMI = mean body mass index ± standard error of the mean, CRP = C-reactive protein, IL = interleukin, TNF = tumour necrosis factor, sTNFR = soluble TNF-α receptor, yr = year, mo = month.

Table 14.2b Summary of published data on the effects of decreased obesity (by weight loss and/or physical activity) on systemic markers of inflammation.

Author group and year of publication	Participants and mean BMI, kg/m^2	Intervention for decreasing obesity	Duration	Magnitude of weight loss	Effects on inflammatory markers
Long-term behavioural changes					
Marfella et al., 2004	67 obese pre-menopausal women BMI 36.5 ± 1.8	1300 kcal/d energy-restricted diet, increased exercise	12 mo	13.4% of weight	CRP ↓ 44%; IL-6 ↓ 62%; TNF-α ↓31%; IL-18 ↓ 30%
Seshadri et al., 2004	78 obese adults, 86% with diabetes or metabolic syndrome BMI 43.5 ± 1.3	≤30 g/d low-carbohydrate diet, or 500 kcal/d deficit energy-restricted diet	6 mo 6 mo	8.5kg 3.5kg	CRP ↓ 12% CRP ↓ 7%
Monzillo et al., 2003	24 obese healthy and diabetic adults BMI 36.7 ± 0.9	500 kcal/d deficit energy-restricted diet and moderate-intensity exercise	26 wk	7.0 % of weight	IL-6 ↓ 41%; no change in TNF-α
Bruun et al., 2003	19 obese men BMI 38.7 ± 0.7	1000–1480 kcal/d energy restricted diet	16 wk	14.7% of weight	IL-6 ↓ 24%; TNF-α ↓ 29%; IL-8 ↓ 30%
Tchernof et al., 2002	25 obese post menopausal women BMI 35.2 ± 1.0	1200 kcal/d American Heart Association Step II diet	14 mo	15.6% of weight, 25% of fat mass, 36,4% of visceral fat	IL-6 ↓ 24%; TNF-α ↓ 29%; IL-8 ↓ 30% CRP ↓ 32%
Ziccardi et al., 2002	56 obese women BMI 37.2 ± 2.2	1300 kcal/d energy-restricted diet, increased exercise	12 mo	12.6% of BMI	IL-6 ↓ 47%; TNF-α ↓ 31%;
Esposito et al., 2002	40 obese women BMI 36.4 ± 2.0	1300 kcal/d energy-restricted diet	12 mo	12.4% of BMI	IL-18 ↓ 41%
Dandona et al., 1998	38 obese women BMI 35.7 ± 0.9	925–1150 kcal/d energy-restricted diet and increased aerobic exercise	1–2 yr	12.3% of weight	TNF-α ↓ 24%

Note: ↓ = decreased, BMI = mean body mass index ± standard error of the mean, CRP = C-reactive protein, IL = interleukin, TNF = tumour necrosis factor, sTNFR = soluble TNF-α receptor, yr = year, mo = month.

Table 14.3 Summary of published data on associations between systemic markers of inflammation and physical activity.

Author group and year of publication (study name)	No. of partic- ipants	Age yr	Association between physical activity and inflammatory markers	Indep. of obesity?
Colbert et al., 2004 (Health, Aging and Body Composition Study)	1507 m 1568 f	70–79	↑ min/wk exercise ⇒ ↓ CRP, IL-6, TNF-α ↑ non-exercise PA ⇒ ↓ CRP, IL-6	Yes, for IL-6 only
Albert et al., 2004 (Pravastatin Inflammation/CRP Evaluation [PRINCE])	1728 m 1105 f	60±12	↑ frequency of PA ⇒ ↓ CRP in men only; no PA-CRP relationship in women	Yes
Jankord and Jemiolo, 2004	12 m	60–74	Very active ⇒ ↓ IL-6; less active ⇒ ↑ IL-10	Not assessed
Pischon et al., 2003 (HPFS and NHSII)	405 m 454 f	40–75 25–42	↑ metabolic equivalent-hours/wk ⇒ ↓ CRP, IL-6, sTNFR1, sTNFR2	No
Reuben et al., 2003 (MacArthur Studies of Successful Aging)	409 m 461 f	70–79	↑ recreational activity ⇒ ↓ CRP, IL-6; ↑ house or yard work ⇒ ↓ CRP, IL-6	Yes
King et al., 2003 (NHANES III)	2036 m 2036 f	>17	Jogging or aerobic dancing >12 time/mo protective for CRP	Yes
Rawson et al., 2003	62 m 47 f	49±12	CRP not related to current physical activity or to physical activity during previous year	No
Wannamethee et al., 2002 (British Regional Heart Study)	3810 m	60–79	↑ volume of PA ⇒ ↓ CRP (no PA ⇒ 2.29 mg/L; vigorous PA ⇒ 1.54mg/L	Yes
Church et al., 2002 (Aerobics Center Longitudinal Study)	722 m	51±10	↑ cardiorespiratory fitness ⇒ ↓ CRP	Yes
Abramson and Vaccarino, 2002 (NHANES III)	1855 m 1783 f	>40	More frequent exercise ⇒ ↓ CRP	Yes
Geffken et al., 2001 (Cardiovascular Health Study)	2473 m 3415 f	≥65	↑ kcal/wk of physical activity ⇒ ↓ CRP	Yes
Taaffe et al., 2000 (MacArthur Studies of Successful Aging)	414 m 466 f	70–79	↑ h/yr moderate-strenuous PA ⇒ ↓ CRP, IL-6	Yes
Milani et al., 2004	161 m 66 f	65.3±11	↑ Cardiorespiratory fitness ⇒ ↓CRP	Yes
Aronson et al., 2004a	562 m 330 f	50±9	↑ Level of fitness ⇒ ↓CRP	Yes
Verdaet et al., 2004	892 m	35–59	No association between leisure time PA and CRP if corrected for BMI	Yes
La Monte et al., 2002	135 f	55±11	↑ Cardiorespiratory fitness ⇒ ↓CRP	Yes

Note: m = male; f = female ↑ = increased, ⇒ led to, HPFS = Health Professionals' Follow-up Study, NHANES III = Third National Health and Nutrition Examination Survey, NHS II = Nurses' Health Study II, PA = physical activity; other symbols and abbreviations as defined in Table 14.2a and Table 14.2b.

3 OBESITY AND THE METABOLIC SYNDROME

According to the World Health Organization (WHO), obesity has reached epidemic proportions globally, with more than one billion overweight adults worldwide – at least 300 million of them obese (WHO, 2003). Obesity can be defined as a disease in which excess body fat has accumulated as a result of a positive energy balance to an extent that health may be adversely affected (ACSM, 2000; WHO, 1998).

Obesity proved to be one of the most essential features of the metabolic syndrome and may be the link that unifies the syndrome (Eckel *et al.*, 2005; Yudkin *et al.*, 2004; Anderson *et al.*, 2001; Yeater, 2000). Overweight and obesity lead to adverse metabolic effects on blood pressure, cholesterol, triglycerides and insulin resistance (WHO, 1998). Shen *et al.* (2003) proposed that obesity may share a synergistic physiologic pathway with insulin resistance, manifested by abnormal metabolism (Carr *et al.*, 2004; Kip *et al.*, 2004). In Figure 14.1 the pathophysiology of cardiovascular disease in the metabolic syndrome originating from central adiposity and innate immunity is illustrated (Reilly and Rader, 2003).

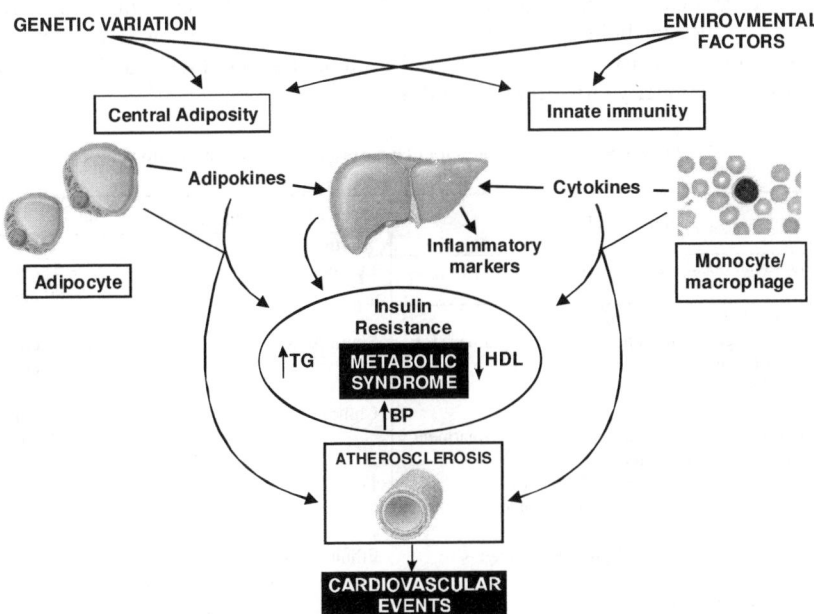

Figure 14.1 Pathophysiology of cardiovascular disease in the metabolic syndrome
Adapted from Reilly and Rader (2003).

Interestingly, although most people with the metabolic syndrome are obese, some of them are of normal body weight. They do however have increased body fat, particularly around the midsection (Debè, 2002). Despres (1997) has shown that visceral fat accumulations greater than 130 cm^2 as measured by computerized tomography or magnetic resonance imaging are strongly associated with the metabolic syndrome. Relative weight gain and obesity in a child from about seven years, can predict the metabolic syndrome in adulthood (Vanhala *et al.*, 1999; Ruderman *et al.*, 1998). Therefore it is not surprising that obesity is a major target of prevention and treatment strategies, given the underlying role of altered adipose tissue metabolism in the metabolic syndrome (Robinson and Graham, 2004), which increases the morbidity and mortality risk of a person (Kip *et al.*, 2004; Ford and Giles, 2003; Isomaa *et al.*, 2001).

According to Reilly and Rader (2003), a major challenge for metabolic research remains the identification of features of adiposity that best reflect increased risk of developing the metabolic syndrome.

4 COMPONENTS OF THE METABOLIC SYNDROME

The Adult Treatment Panel (ATP III) identified six components of the metabolic syndrome that relate to CHD namely, abdominal obesity, atherogenic dyslipidemia, raised blood pressure, insulin resistance/glucose intolerance, pro-inflammatory state and prothrombotic state (Alberti *et al.*, 2005; Grundy *et al.*, 2004). Recently the International Diabetes Federation (IDF) proposed new guidelines for diagnosing the metabolic syndrome (Table 14.4). Central to the new IDF definition is abdominal obesity, identified by waist circumference (WC) and classified according to ethnicity (Alberti *et al.*, 2005).

Table 14.4 New IDF guidelines for diagnosis of the Metabolic Syndrome.

IDF guidelines	Ethnic group-specific WC values		
	Ethnic group	Gender	WC
Abdominal obesity defined by WC	European	Male	= 94 cm
		Female	= 80 cm
Plus another two of the following:			
Raised TG concentrations or specific treatment for this lipid abnormality	South Asian	Male	= 90 cm
▪ >150 mg/dL (1.7 mmol/L)		Female	= 80 cm
	Chinese	Male	= 90 cm
Reduced HDL concentrations or specific treatment for this lipid abnormality		Female	= 80 cm
▪ <40 mg/dL (1.03 mmol/L) Males	Japanese	Male	= 85 cm
▪ <50 mg/dL (1.29 mmol/L) Females		Female	= 90 cm
Raised BP or treatment of previously diagnosed hypertension	Ethnic South and Central American	Male	*
		Female	*
▪ ≥130 / 85 mmHg			
	Sub-Saharan African Eastern	Male	**
Raised FPG concentration or previously diagnosed type 2 diabetes		Female	**
▪ ≥100 mg/dL (5.6 mmol/L)			
	Mediterranean Middle East	Male	**
		Female	**

Note: BP = Blood pressure; WC = Waist circumference; TG = Triglyceride; FPG = Fasting plasma glucose; HDL = High-density lipoprotein; *Use South Asian recommendations until specific data are available; **Use European recommendations until specific data are available.
Adapted from Alberti *et al.*, (2005).

5 OBESITY AS AN INFLAMMATORY CONDITION

Das (2001) proposed that obesity could be an inflammatory disorder due to low-grade systemic inflammation, and this view is supported by a vast number of researchers (Dandona *et al.*, 2004, 2005; Nawrocki and Scherer, 2004; Ramos *et al.*, 2003; Weyer *et al.*, 2002). CRP is a highly conserved molecule and a member of the pentraxin family of proteins (Du Clos, 2000). CRP is an acute phase reactant, synthesized primarily in hepatocytes and secreted by the liver in response to a variety of inflammatory cytokines of which IL-6 and TNF-α are the main ones involved. CRP increases rapidly in response to trauma, inflammation and infection and decreases just as rapidly with the resolution of the condition (Heilbronn and Clifton, 2002; Das, 2001). Thus, the measurement of CRP can be used to monitor inflammatory states. According to Das (2001), CRP has a role in the function of the innate immune system: it activates complement, binds to Fc (constant fragment of an antibody molecule) receptors and acts as an opsonin for various pathogens. Binding of CRP to Fc receptors leads to the generation of pro-inflammatory cytokines. CRP can recognize self- and foreign molecules based on pattern recognition. Thus enhanced levels of CRP can be used as a marker of inflammation (Das, 2001).

Several markers of inflammation correlate with insulin action and indirect measures of adiposity, especially fat distribution or patterning, and provide a possible link between obesity and insulin resistance (Weyer *et al.*, 2002; Pannacciulli *et al.*, 2001). The synthesis of adipose tissue TNF-α could induce the production of IL-6, CRP and other acute phase reactants (Yudkin *et al.*, 1999), therefore contributing to the maintenance of a chronic low-grade inflammation state involved in the progression of obesity and its associated co-morbidities.

Many patients with the metabolic syndrome have a state of chronic low-grade inflammation indicated by high CRP levels, which may increase their risk for future adverse and cardiovascular events (Ford, 2003), even in apparently healthy individuals (Wang *et al.*, 2002).

5.1 Measures of adiposity

Clinically the prevalence of overweight and obesity is assessed by using body mass index (BMI), defined as the weight in kilograms divided by the square of the height in meters (kg/m^2) (Norton and Olds, 1996). Overweight as defined by the WHO is a BMI higher than 25 kg/m^2 and obesity as a BMI of higher than 30 kg/m^2 (Grundy *et al.*, 1999). It should be noted however that although BMI is widely used in the literature to measure degree of overweight and obesity, it is not a measure of fatness, but rather a measure of heaviness instead (Heyward and Wagner, 2004; WHO, 2000). For this reason many studies proposed that the definition of obesity based on height and weight needs to be modified as cited by Ruderman *et al.* (1998). Waist-to-hip ratio (WHR), waist circumference (WC) and percentage body fat (% BF) could therefore be more accurate indicators of obesity. Increased WC is an indicator of abdominal obesity (Grundy *et al.*, 2004), and predicts greater cardiovascular risk better (Charlton *et al.*, 2001; Grundy *et al.*, 1999). It is however unclear in most studies which definition of WC was used, since there are three common definitions of WC (umbilicus, visible narrowing, halfway between iliac crest and costal border).

Rexrode *et al.* (2003) found BMI and WC to be strongly correlated with CRP and IL-6. The correlations of BMI with WC were 0.73 and age-adjusted CRP with BMI and WC were 0.42 (p = 0.0001) and 0.33 (p = 0.0001) respectively. Similarly, IL-6 were 0.33 (p = 0.0001) for BMI and 0.29 (p = 0.0001) for WC. In particular, CRP and IL-6 as markers of inflammation will be considered in this review.

5.2 Relationship between CRP and measures of adiposity

In the literature search, 39 studies were identified which met the inclusion criteria. The studies are summarized in Tables 14.1–14.3. The studies were undertaken between 1998 and 2004. In Table 14.2, there were approximately 4254 participants, the majority being women. The ages varied from 14 to 63 years. Participants were mostly characterized as healthy. The mean BMI was 27.8 ± 6.5. All of the studies measured BMI in relationship to CRP. Six studies evaluated abdominal obesity by taking waist circumference measures (Slabbert, 2004; Barinas-Mitchell *et al.*, 2001; Heilbronn *et al.*, 2001; Lemieux *et al.*, 2001; Vozarova *et al.*, 2001; Hak *et al.*, 1999). The measures were not uniform across all samples. [Slabbert (2004), Heilbronn *et al.* (2001) and Barinas-Mitchell *et al.* (2001) measured at the narrowest point between the lower costal and iliac crest; Vazarova *et al.* (2001) measured at the umbilicus in the supine position; neither Lemieux *et al.* (2001) nor Hak *et al.* (1999) specified where they measured.] Nevertheless, all six studies reported WC as a strong indicator of abdominal obesity. WHR was measured in six studies (Slabbert, 2004; Vikram *et al.*, 2003; Tchernof *et al.*, 2002; Ziccardi *et al.*, 2002; Barinas-Mitchell *et al.*, 2001; Chambers *et al.*, 2001). An indication of the relationship between % BF and markers of inflammation was given by three studies (Vikram *et al.*, 2003; Weyer *et al.*, 2002; Barinas-Mitchell *et al.*, 2001). All of the above found a positive relationship between markers of inflammation (CRP and IL-6), total obesity (BMI) and abdominal obesity (WC and WHR). Subjects with high WHR had higher median levels of CRP (p = 0.001), CRP correlated significantly with WHR, p = 0.02) after adjustment for age in young adults and adolescents of North India (Vikram *et al.*, 2003).

Evidence is strong that circulating levels of inflammatory markers are elevated with total and abdominal obesity (see Table 14.1). This is possibly owing to a higher secretion rate of cytokines by adipose tissue in obese people (Nicklas *et al.*, 2005). The relationship between CRP and abdominal obesity showed strong associations with WC, WHR and visceral fat as reported by Vikram *et al.* (2003), Lemieux *et al.* (2001) and Slabbert (2004). Among body fatness and fat distribution indices, total body fat mass was the variable that showed the highest correlation with CRP levels (r = 0.41, p < 0.0001) (Lemieux *et al.*, 2001). In cross-sectional studies, elevated CRP levels have been associated with proxy indicators of elevated body fatness (body weight and BMI) (Tchernof *et al.*, 2002). In studies by Saijo *et al.* (2004) and Forouhi *et al.* (2001) the amount of visceral fat was a better determinant of CRP levels than other levels of obesity, including fat mass. The location of the body fat, independent of the total amount, is therefore an important factor affecting chronic inflammation. There is some evidence from observational studies involving both men and women, that in addition to total body fat, visceral or abdominal body fat (measured by WC and WHR) may be an independent predictor of inflammatory markers (Saijo *et al.*, 2004; Sites *et al.*, 2002; Ziccardi *et al.*, 2002; Forouhi *et al.*, 2001; Lemieux *et al.*, 2001; Bertin *et al.*, 2000; Barinas-Mitchell *et al.*, 2001).

Adipose tissue is known to be a secretory organ producing cytokines, acute phase reactants and other circulating factors (Kershaw and Flier, 2004; Trayhurn and Wood, 2004; Ouchi *et al.*, 2003; Wong *et al.*, 2003). These "adipokines" are mostly not produced by the adipocyte itself but by the infiltration of macrophages into the adipocytes (Fain *et al.*, 2004; Weisberg *et al.*, 2003). Research by Yudkin *et al.* (1999) suggests that the cytokines, arising in part from adipose tissue, might be partly responsible for the metabolic, hemodynamic, and hemostatic abnormalities that cluster with insulin resistance. One of the determinants of CRP is IL-6, which is secreted by adipocytes (Papanicolaou *et al.*, 1998; Mohamed-Ali *et al.*, 1997). In vivo release of IL-6 and TNF-soluble receptors from subcutaneous abdominal adipose tissue has been shown to correlate with BMI and body-fat proportion (Mohamed-Ali *et al.*, 1999; Yudkin *et al.*, 1999). In one in-vitro study TNF-α release from abdominal subcutaneous adipose tissue was 7.5 fold higher in tissue from obese (BMI 30–40 kg/m^2) than lean (BMI < 25 kg/m^2) subjects (Kern *et al.*, 1995). Ouchi and associates (2003) reported not only that CRP is expressed in adipose tissue, but also that CRP and adiponectin mRNA levels are highly inversely related. (Adiponectin is a protein with anti-inflammatory properties.) Thus, both in vivo and in vitro studies confirm that adipose tissue expression and release of cytokines are elevated in people with a higher adipose mass.

It is however interesting that most studies did not measure abdominal obesity, but only BMI as an indication of obesity which does not give any indication of a person's body composition (Vikram *et al.*, 2003; Weyer *et al.*, 2002). Strong associations of CRP and BMI were found in obese individuals (Aronson *et al.*, 2004b; Rexrode *et al.*, 2003; Ridker, 2003; Vikram *et al.*, 2003; Lemieux *et al.*, 2001; Hak *et al.*, 1999), but many studies do not differentiate between android and gynoid type obesity (abdominal obesity). This complicates the interpretation of the CRP relationship to abdominal obesity and could lead to confusing assumptions in this regard. The relationship between CRP and BMI in the normal weight range has not been well characterized (Rexrode *et al.*, 2003). Rexrode and associates (2003) could not determine whether obesity causes elevated CRP levels directly, or whether higher CRP levels are a marker of other intermediate conditions such as atherosclerosis or insulin resistance which influence the underlying burden of inflammation among overweight and obese individuals.

The contribution of the association between abdominal obesity and metabolic risk remains the subject of considerable debate (Frayne, 2000). There are however studies that indicate that adipocytes within the deep visceral compartment are more metabolically active than adipocytes within the superficial visceral compartment and thus a stronger predictor of insulin resistance (Wong *et al.*, 2003; Carey, 1997).

6 THE INFLUENCE OF PHYSICAL ACTIVITY AND WEIGHT LOSS ON SYSTEMIC INFLAMMATORY MARKERS ASSOCIATED WITH OBESITY

The etiology of obesity represents a complex interaction of genetics, diet, metabolism and physical activity levels (Das, 2001). Diet and physical activity play very significant roles in the prevalence of obesity. In addition to the consumption of high-energy food, physical activity is a key factor in the energy balance equation (Epstein *et al.*, 2000).

6.1 Weight loss and inflammatory markers associated with obesity

Evidence from intervention studies showed that weight loss leads to a reduction in the state of chronic systemic inflammation, which is associated with increased adipose tissue mass in overweight and obese subjects (summarized in Table 14.2). Weight loss by means of dietary restriction (short term) (five studies), long term behavioural changes (eight studies) and randomized controlled trails (two studies) are summarized. Intense short term dietary restrictions of 3–12 weeks were indicated. Of the five studies summarized, two studies led to decreased CRP (Xydakis *et al.*, 2004; Heilbronn *et al.*, 2001). Two studies showed decreased IL-6 (Gallistl *et al.*, 2001; Bastard *et al.*, 2000a). No change was found in TNF-α (Xydakis *et al.*, 2004; Bastard *et al.*, 2000a), CRP (Bastard *et al.*, 2000a) or sTNFR2 (Bastard *et al.*, 2000b). Long term behavioural changes indicated significant reductions in CRP in four of the studies, reductions in IL-6 and TNF-α in five studies and reduced IL-18 and IL-8 was both found in two studies.

Several markers of inflammation, including CRP, IL-6, TNF-α and TNF-α receptors are reduced after weight loss achieved through short-term intense dietary restriction (Xydakis *et al.*, 2004; Gallistl *et al.*, 2001; Heilbronn *et al.*, 2001; Bastard *et al.*, 2000a,b). Most of the dietary weight-loss studies showed the magnitude of decrease in inflammatory markers to be linearly related to the amount of weight lost. An example is when CRP concentrations were reduced from 3.1 ± 0.7 mg/l to 1.6 ± 0.8 mg/l in a study of postmenopausal women after a 14 month individualized weight-loss program by Tchernof and co-authors (2002). The reductions in CRP reportedly correlated positively with changes in both body weight and fat mass. The percentage change for the weight loss effect for weight and fat mass was –15.5% and –25% respectively ($p < 0.0001$) for CRP after a weight loss period (Tchernof *et al.*, 2002). Decreases in CRP, IL-6 and TNF-α in a group of pre-menopausal women after 10% weight reduction correlated with changes in BMI (TNF-α, $r = 0.35$, $p < 0.05$ and IL-6, $r = 0.31$, $p < 0.05$) but were more strongly related to changes in waist-hip-ratio (TNF-α, $r = 0.54$, $p < 0.01$ and IL-6, $r = 0.45$, $p < 0.01$) (Marfella *et al.*, 2004; Ziccardi *et al.*, 2002). In other studies, weight loss reduced CRP levels from 5.0 ± 0.5 mg/l to 4.3 ± 0.5 mg/l in women with metabolic syndrome (Xydakis *et al.*, 2004), and IL-6 concentrations from 2.75 ± 1.51 to 2.3 ± 0.91 pg/ml in women with insulin resistance (Monzillo *et al.*, 2003). Heilbronn *et al.* (2001) reported that CRP values decreased 26% with a 19% weight-loss, and decreased 32%–34% with a 15%–16% weight-loss (Esposito *et al.*, 2003; Tchernof *et al.*, 2002).

The results of the current review suggest that there may be a dose-response effect between the degree of weight loss and its capacity to attenuate chronic inflammation. Longitudinal studies are needed to determine whether a reduced incidence of cardiovascular disease and diabetes is associated with the decline in CRP concentrations seen with weight loss (Nicklas *et al.*, 2005).

6.1.1 Mechanism of effect

One of the postulated mechanisms by which weight loss reduces circulating markers of inflammation is through a decrease in adipose-tissue cytokine production (Nicklas *et al.*, 2005). In Figure 14.2 the possible mechanisms by which weight loss and exercise training reduce sources of inflammation that lead to chronic activation of a pro-inflammatory state are illustrated. According to Nicklas *et al.* (2005) weight loss and increased activity affect the immune system by reducing the number of mononuclear cells in the peripheral blood, which are a source of pro-inflammatory cytokines (such IL-6 and TNF-α and its receptors). A reduction in adipose tissue would not only reduce the volume of adipocytes and pre-adipocytes, but also decrease the number of endothelial cells and macrophages that reside there. These cells produce many pro-inflammatory mediators such as CRP, serum amyloid protein A (SAA) and cytokines. Weight loss and exercise may also increase the expression of anti-inflammatory mediators such as IL-10 and IL-1 receptor antagonist (IL-1ra) in cells. The resulting circulatory changes could, in turn, cause the liver to contribute by decreasing its production of fibrinogen and other pro-inflammatory mediators.

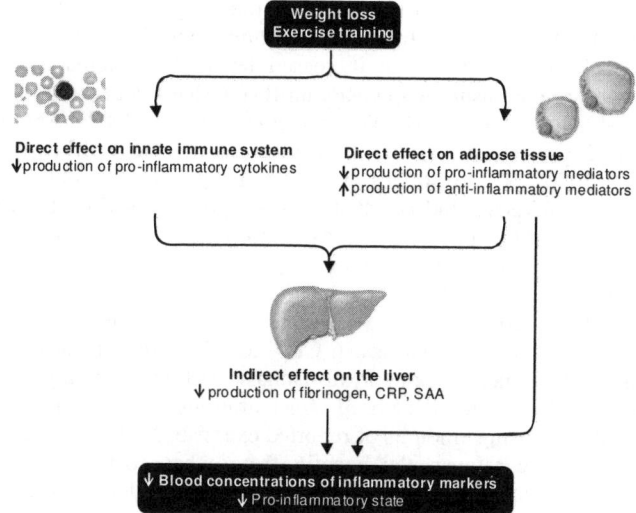

Figure 14.2 Mechanisms by which weight loss and exercise training reduce chronic low-grade inflammation (Adapted from Nicklas *et al.* (2005)).

Grundy *et al.* (2004) reported that physical activity is a modality associated with successful weight reduction, particularly for weight maintenance. The studies in Table 14.3 revealed that dietary weight loss and exercise are likely more effective than weight reduction alone in reducing inflammation.

6.2 ﾟhysical activity associations with CRP, IL-6 and obesity

Physical activity proved effective in lowering measures of adiposity (BMI, WHR, WC and percentage body fat) (Milani *et al.*, 2004; Verdaet *et al.*, 2004; Pischon *et al.*, 2003; Tchernof *et al.*, 2002) and obesity related inflammatory markers (CRP and IL-6) (Aronson *et al.*, 2004a,b; Church *et al.*, 2002), thereby indicating a potential anti-inflammatory effect of physical activity (Wannamethee *et al.*, 2002; La Monte *et al.*, 2002). According to Aronson *et al.* (2004a) the importance of physical activity in improving the pro-inflammatory state associated with the metabolic syndrome needs to be emphasized because it could be more efficient than the use of medication.

A summary of 16 published studies on the associations between systemic markers of inflammation and physical activity is shown in Table 14.3. Physical activity and fitness demonstrated an inverse relationship with CRP and IL-6 in 12 of the studies. Most studies kept the significant inverse relationship between physical activity and CRP / IL-6 even when corrected for obesity (by measure of BMI or WC). Two studies found no association between physical activity and CRP (Verdaet *et al.*, 2004; Rawson *et al.*, 2003).

Increased leisure time physical activity (recreational, house and yard work) indicated reductions in CRP and IL-6 in a study by Reuben *et al.* (2003), and this association was independent of obesity. However, Verdaet *et al.* (2004) found no association between physical activity and inflammatory markers when CRP was corrected for BMI. Increased frequency, volume and intensity (moderate to strenuous) of physical activity and increased level of cardiorespiratory fitness showed an inverse relationship with CRP and IL-6 (Albert *et al.*, 2004; Aronson *et al.*, 2004a; Milani *et al.*, 2004; Wannamethee *et al.*, 2002; Abramson and Vaccarino, 2002; Church *et al.*, 2002; La Monte *et al.*, 2002; Geffken *et al.*, 2001; Taaffe *et al.*, 2000). These studies were all independent of obesity. Clearly, evidence for an inverse, independent dose-response relation between CRP concentration and level of physical activity in both men and women are provided. This relation did not seem to alter with age as is evident in studies by Colbert *et al.* (2004) and Geffken *et al.* (2001).

Pischon *et al.* (2003) assessed physical activity by metabolic equivalent-hours per week and found reductions in CRP and IL-6, dependent of obesity, in 405 men and 454 women (Health Professionals' Follow-up Study and Nurses' Health Study II). Trends for decreased concentrations of inflammatory markers were linear with increasing amounts of reported exercise (Nicklas *et al.*, 2005).

The health and cardiovascular benefits from enhanced fitness may have an anti-inflammatory mechanism and may at least be short-term effects mediated through the already mentioned mechanisms (Wannamethee *et al.*, 2002; La Monte *et al.*, 2002).

7 CONCLUSION

Data from several large population-based cohorts showed an inverse association between markers of systemic inflammation and physical activity or fitness status. Small-scale intervention studies also proved that exercise training diminishes inflammation associated with obesity. To date, data from randomized controlled trials designed to definitively test the effects of weight loss or exercise training, or both, on inflammation are limited. Future studies are required to define the amount of weight loss needed for clinically meaningful reductions of inflammation. In addition, controlled studies are necessary to clarify the effect of exercise training on chronic systemic inflammation. The mechanisms by which weight loss and increased physical activity reduce inflammation have yet to be elucidated.

8 ACKNOWLEDGEMENTS

The financial assistance of the National Research Foundation (NRF, South Africa) towards this research is hereby acknowledged. Opinions expressed and conclusions arrived at, are those of the author and are not necessarily to be attributed to the NRF.

9 REFERENCES

Abramson, J.L. and Vaccarino, V., 2002, Relationship between physical activity and inflammation among apparently healthy middle-aged and older US adults. *Archives of Internal Medicine*, **162**, 1286–1292.

Alberti, G., Zimmet, P. and Lefebvre, P., 2005, Metabolic syndrome: Introducing new globally applicable guidelines. IDF task force on epidemiology. [Web:] www.idf.org/webdata/docs/IDF_Metasyndrome_definition.pdf [Date of access: 16 Oct. 2005].

Albert, M.A., Glynn, R.J. and Ridker, P.M., 2004, Effect of physical activity on serum C-reactive protein. *American Journal of Cardiology*, **93**, 221–225.

Anderson, P.J., Critchley, J.A.J.H., Chan, J.C.N., Cockram, C.S., Lee, Z.S.K., Thomas, G.N. and Tomlinson, B., 2001, Factor analysis of the metabolic syndrome: Obesity versus insulin resistance as the central abnormality. *International Journal of Obesity*, **25**, 1782–1788.

American College of Sports Medicine (ACSM), 2000, ACSM'S guidelines for exercise testing and prescription 6th ed. (Philadelphia : Lippencott Williams and Wilkens) 368 p.

Appel, S.J., Giger, J.N. and Floyd, N.A., 2004, Dysmetabolic syndrome: Reducing cardiovascular risk. *The Nurse Practitioner*, **29**(10), 18–35.

Aronson, D., Sella, R., Sheikh-Ahmad, M., Kerner, A., Avizohar, O., Rispler, S., Barha, P., Markiewicz, W., Levy, Y. and Brook, G.J., 2004a, The association between cardiorespiratory fitness and C-reactive protein in subjects with the metabolic syndrome. *Journal of the American College of Cardiology*, **44**(10), 2003–2007.

Aronson, D., Sheikh-Ahmad, M., Avizohar, O., Kerner, A., Sella, R., Bartha, P., Markiewicz, W., Levy, Y. and Brook, G.J., 2004b, C-Reactive protein is inversely related to physical fitness in middle-aged subjects. *Atherosclerosis*, **176**, 173–179.

Barinas-Mitchell, E., Cushman, M., Meihlan, E.N., Tracy, R.P. and Kuller, L.H., 2001, Serum levels of C-reactive protein are associated with obesity, weight gain and hormone replacement therapy in healthy post-menopausal women. *American Journal of Epidemiology*,**153**, 1094–1101.

Bastard, J.P., Jardel, C., Bruckert, E., Blondy, P., Capeau, J. and Laville, M., 2000a, Elevated levels of interleukin-6 are reduced in serum and subcutaneous adipose tissue of obese women after weight loss. *Journal of Clinical Endocrinology and Metabolism*, **85**, 3338–3342.

Bastard, J.P., Jardel, C., Bruckert, E., Vidal, H. and Hainque, B., 2000b, Variations in plasma soluble tumour necrosis factor receptors after diet-induced weight loss in obesity. *Diabetes, Obesity and Metabolism*, **2**, 323–325.

Bertin, E., Nguyen, P., Guenounou, M., Durlach, V., Potron, G. and Leutenegger, M., 2000, Plasma levels of tumour necrosis-alpha are essentially dependant on visceral fat amount in type 2 diabetic patients. *Diabetes and Metabolism*, **26**, 178–182.

Bouchard, C. and Blair, S.N., 1999, Introductory comments for the consensus on physical activity and obesity. *Medicine and Science in Sports and Exercise*, S499–S501.

Bruun, J.M., Verdich, C., Toubro, A., Astrup, A. and Richelsen, B., 2003, Association between measures of insulin sensitivity and circulating levels of interleukin-8, interleukin-6 and tumour necrosis factor-alpha. Effect on weight loss in obese men. *European Journal of Endocrinology*, **148**, 535–542.

Carr, D.B., Utzschneider, K.M., Hull, R.L., Kodama, K., Retzlaff, B.M., Brunzell, J.D., Shofer, F.B., Fish, B.E., Knopp, R.H. and Kahn, S.E., 2004, Intra-abdominal fat is a major determinant of the National Cholesterol Education Program Panel III criteria for the metabolic syndrome. *Diabetes*, **53**, 2087–2094.

Carey, G.B., 1997, The swine as a model for studying exercise-induced changes in lipid metabolism. *Medicine and Science in Sports and Exercise*, **11**, 1437–1443.

Chambers, J.C., Eda, S., Bassett, P., Karim, Y., Thompson, S.G., Gallimore, J.R., Pepys, M.B. and Kooner, J.S., 2001, C-Reactive protein, insulin resistance, central obesity, and coronary heart disease risk in Indian Asians from the United Kingdom compared with European whites. *Circulation*, **104**, 145–150.

Charlton, K.E., Schloss, I., Visser, M., Lambert, E.V., Kolbe, T., Levitt, N.S. and Temple, N., 2001, Waist circumference predicts clustering of cardiovascular risk factors in older South Africans. *Cardiovascular Journal of South Africa*, **12**(3), 142–150.

Church, T.S., Barlow, C.E., Earnest, C.P., Kampert, J.B., Priest, E.L. and Blair, S.N., 2002, Associations between cardiorespiratory fitness and C-reactive protein in men. *Arteriosclerosis, Thrombosis and Vascular Biology*, **22**, 1869–1876.

Colbert, L.H., Visser, M., Simonsick, E.M., Tracy, R.P., Newman A.B. and Kritchevsky, S.B., 2004, Physical activity, exercise, and inflammatory markers in older adults: Findings from the health, aging and body composition study. *Journal of American Geriatrics Society*, **52**, 1098–1104.

Dandona, P., Aljada, A. and Bandyopadhyay, A., 2004, Inflammation: The link between insulin resistance, obesity and diabetes. *Trends in Immunology*, **25**(1), 4–7.

Dandona, P., Aljada, A., Chaudhuri, A., Mohanty, P. and Garg, R., 2005, Metabolic syndrome: A comprehensive perspective based on interactions between obesity, diabetes, and inflammation. *Circulation*, **111**, 1448–1454.

Dandona, P., Weinstock, R., Thusu, K., Abdel-Rahman, E., Aljada, A. and Wadden, T., 1998, Tumour necrosis factor-alpha in sera of obese patients: Fall with weight loss. *Journal of Clinical Endocrinology and Metabolism*, **83**, 2907–2910.

Das, U.N., 2001, Is obesity an inflammatory condition? *Nutrition*, **17**, 953–966.

Debè, J.R., 2002, Reversing the number 1 cause of illness, obesity and accelerated aging. [Web:] http://www.drdebe.com/MetabolicSyndrome.htm [Date of access: 20 March 2005].

Deen, D., 2004, Metabolic syndrome: Time for action. *American Family Physician*, **69**(12), 2875–2882.

Despres, J.P., 1997, Vesical obesity, insulin resistance and dyslipidemic: Contributive of endurance exercise training to the treatment of the pleumetabolic syndrome. *Exercise and Sport Science Reviews*, **25**, 271–300.

Du Clos, T.W., 2000, Function of C-reactive protein. *Annals of Medicine*, **32**, 274.

Eckel, R.H., Grundy, S.M. and Zimmet, P.Z., 2005, The metabolic syndrome. *Lancet*, **365**, 1415–1428.

Epstein, L.H., Paluch, R.A., Gorgy, C.C. and Dorn, J., 2000, Decreasing sedentary behaviors in treating pediatric obesity. *Archives of Pediatric and Adolescent Medicine*, **154**, p. 220.

Esposito, K., Pontillo, A., Ciotola, M., Di Palo, C., Grella, E. and Nicoletti, G., 2002, Weight loss reduces interleukin-18 levels in obese women. *Journal of Clinical Endocrinology and Metabolism*, **87**, 3864–3866.

Esposito, K., Pontillo, A., Di Palo, C., Giugliano, G., Masella, M., Marfella, R. and Giugliano, D., 2003, The effect of weight loss and lifestyle changes on vascular inflammatory markers in obese women: A randomized trial. *Journal of the American Medical Association*, **289**, 1799–1804.

Fain, J.N., Madan, A.K., Hiler, M.L., Cheerma, P. and Bahouth, S.W., 2004, Comparison of the release of adipokines by adipose tissue, adipose tissue matrix and adipoytes from visceral and subcutaneous abdominal adipose tissue of obese humans. *Endocrinology*, **145**, 2273–2282.

Ford, E.S., 2003, The metabolic syndrome and C-reactive protein, fibrinogen, and leukocyte count: Findings from the third national health and nutrition examination survey. *Atherosclerosis*, **168**, 351–358.

Ford, E.S. and Giles, W.H., 2003, A comparison of the prevalence of the metabolic Syndrome using two proposed definitions. *Diabetes Care*, **26**(3), 575–581.

Forouhi, N.G., Sattar, N. and McKeigue, P.M., 2001, Relation of C-reactive protein to body fat distribution and features of the metabolic syndrome in Europeans and South Asians. *International Journal of Obesity Related Disorders*, **25**, 1327–1331.

Frayne, M.J., 2000, Visceral fat and insulin resistance: Causative or correlative? *British Journal of Nutrition*, **83**, S71–S77.

Gallistl, S., Sudi, K.M., Aigner, R. and Borkenstein, M., 2001, Changes in serum interleukin-6 concentrations in obese children and adolescents during a weight reduction program. *International Journal of Obesity Related Metabolic Disorders*, **25**, 1640–1643.

Geffken, D., Cushman, M., Burke, G., Polak, J., Sakkinen, P. and Tracy, R., 2001, Association between physical activity and markers of inflammation in a healthy elderly population. *American Journal of Epidemiology*, **153**, 242–250.

Grundy, S.M., Blackburn, C., Higgins, M., Lauer, R., Perri, M.G. and Ryan, D., 1999, Physical activity in the prevention and treatment of obesity and its co-morbidities. *Medicine and Science in Sports and Exercise*, S504–S508.

Grundy, S.M., Brewer, B., Cleeman, J.I., Smith, S.C. and Lenfant, C., 2004, Definition of metabolic syndrome: Report of the national heart, lung, and blood institute/ American heart association conference on scientific issues related to definition. *Circulation*, **109**, 433–438.

Hak, A.E., Stehouwer, C.D.A., Bots, M.L., Polderman, K.H., Schalkwijk, C.G., Westendorp, I.C.D., Hoffman, A. and Witteman, J.C.M., 1999, Associations of C-reactive protein with measures of obesity, insulin resistance, and subclinical atherosclerosis in healthy, middle-aged women. *Arteriosclerosis, Thrombosis and Vascular Biology*, **19**, 1986–1991.

Heilbronn, L.K. and Clifton, R.M., 2002, C-reactive protein and coronary artery disease: Influence of Obesity, Caloric Restriction and Weight Loss. *Journal of Nutritional Biochemistry*, **13**, 316–322.

Heilbronn, L.K., Noakes, M. and Clifton, P.M., 2001, Energy restriction and weight loss on very-low-fat diets reduce C-reactive protein concentrations in obese, healthy women. *Arteriosclerosis, Thrombosis and Vascular Biology*, **21**, 968–970.

Heyward, V.H. and Wagner, D.R., 2004, Applied body composition assessment. (Illinois : Human Kinetics), 268 p.

Isomaa, B., Algren, P., Tuomi, T., Forsen, B., Lahti, K., Nissen, M., Taskinen, M.R. and Groop, L., 2001, Cardiovascular morbidity and mortality associated with the metabolic syndrome. *Diabetes care*, **24**(4), 683–689.

Jankord, R. and Jemiolo, B., 2004, Influence of physical activity on serum IL-6 and IL-10 levels in healthy older men. *Medicine and Science in Sports and Exercise*, **36**, 960–964.

Kern, P., Saghizadeh, M., Ong, J., Bosch, R., Deem, R. and Simsolo, R., 1995, The expression of tumour necrosis factor in human adipose tissue. Regulation by obesity, weight loss and relationship to lipoprotein lipase. *Journal of Clinical Investigation*, **95**(5), 2111–2119.

Kershaw, E.E. and Flier, J.S., 2004, Adipose tissue as an endocrine organ. *Journal of Clinical Endocrinology and Metabolism*, **89**, 2548–2556.

King, D.E., Carek, P., Mainous, A.G. III. and Pearson, W.S., 2003, Inflammatory markers and exercise: Differences related to exercise type. *Medicine and Science in Sport and Exercise*, **35**, 575–581.

Kip, K.E., Marroquin, O.C., Kelley, D.E., Johnson, B.D., Kelsey, S.F., Shaw, L.J., Rogers, W.J. and Reis, S.E., 2004, Clinical importance of obesity versus the metabolic syndrome in cardiovascular risk in women: A report from the

Women's Ischemia Syndrome Evaluation (WISE) Study. *Circulation*, **109**, 706–713.

Lam, K.S.L., Xu, A., Wat, N.M.S., Tso, A.W.K. and Ip, M.S.M., 2003, Obesity as the key player in the metabolic syndrome. *International Congress Series*, **1262**, 542–545.

La Monte, M.J., Durstine, L.J., Yanowitz, F.G., Lim, T., Dubose, K.D., Davis, P. and Ainsworth, B.E., 2002, Cardiorespiratory fitness and C-reactive protein among a tri-ethnic sample of women. *Circulation*, **106**, 403–406.

Lemieux, I., Pascot, A., Prud'Homme, D., Alméras, N., Bogaty, P., Nadeau, A., Bergeron, J. and Deprés, J., 2001, Elevated C-reactive protein: Another component of the atherothrombotic profile of abdominal obesity. *Arteriosclerosis, Thrombosis and Vascular Biology*, **21**, 961–967.

Leon, A.S., 1997, Physical activity and cardiovascular health: A national consensus. (Illinois : Human Kinetics), 272 p.

Marfella, R., Esposito, K., Siniscalshi, M., Cacciapuoti, F. and Labriola, D., 2004, Effect of weight loss on cardiac synchronization and pro-inflammatory cytokines in pre-menopausal obese women. *Diabetes Care*, **27**, 47–52.

Milani, R.V., Lavie, C.J. and Mehra, M.R., 2004, Reduction in C-reactive protein through cardiac rehabilitation and exercise training. *Journal of the American College of Cardiology*, **43**(6), 1056–1061.

Mohamed-Ali, V., Goodrick, S., Bulmer, K., Holly, J.M., Yudkin, J.S. and Coppack, S.W., 1999, Production of soluble tumour-necrosis factor receptors by human subcutaneous adipose tissue in vivo. *American Journal of Physiology*, 277, E971–E975.

Mohamed-Ali, V., Goodrick, S., Rawesh, A., Katz, D.R., Miles, J.M. and Yudkin, J.S., 1997, Subcutaneous adipose tissue releases interleukin-6, but not tumour necrosis factor-alpha, in vivo. *Journal of Endocrinology and Metabolism*, **82**, 4196–4200.

Monzillo, L.U., Hamdy, O., Horton, E.S., Ledbury, S., Mullooly, C. and Jarema, C., 2003, Effect of lifestyle modification on adipokine levels in obese subjects with insulin resistance. *Obesity Research*, **11**, 1048–1054.

Nawrocki, A.R. and Scherer, P.E., 2004, The delicate balance between fat and muscle: Adipokines in metabolic disease and musculoskeletal inflammation. *Current Opinion in Pharmacology*, **4**, 281–289.

Norton, K. and Olds, T., 1996, Anthropometrica. A text book of body measurement for sports and health courses. (Sydney: University of New South Wales Press.) 411 p.

Nicklas, B.M., Ambrosius, S.P., Miller, G.D., Penninx, B.W. and Loeser, R.F., 2004, Diet-induced weight loss, exercise and chronic inflammation in older, obese adults: a randomized controlled clinical trial. *American Journal of Clinical Nutrition*, **79**, 544–551.

Nicklas, B.J., Tongjian, Y. and Pahor, M., 2005, Behavioral treatments for chronic systemic inflammation: Effects of dietary weight loss and exercise training. *Canadian Medical Association Journal*, **172**(9), 1199–1209.

Ouchi, N., Kihara, S., Funahashi, T., Nakamura, T. and Kumuda, M., 2003, Reciprocal association of C-reactive protein with adiponectin in blood stream and adipose tissue. *Circulation*, **107**, 671–674.

Pannacciulli, N., Cantatore, F.P., Minenna, A., Bellacicco, M., Giorgino, R. and De Pergola, G., 2001, C-reactive protein is independently associated with total body fat, central fat, and insulin resistance in adult women. *International Journal of Obesity*, **25**(10), 1416–1420.

Papanicolaou, D.A., Wilder, R.L., Manolagas, S.C. and Chrousos, G.P., 1998, The pathologic roles of interleukin-6 in human disease. *Annals of Internal Medicine*, **128**, 127–137.

Pischon, T., Hankinson, S.E., Hotamisligil, G.S., Rifai, N. and Rimm, E.B., 2003, Leisure-time physical activity and reduced plasma levels of obesity-related inflammatory markers. *Obesity Research*, **11**(9), 1055–1064.

Ramos, E.J.B., XU, Y., Romanova, I., Middleton, F., Chen, C., Quinn, R., Inui, A., Das, U. and Meguid, M.M., 2003, Is obesity an inflammatory disease? *Surgery*, **134**(2), 329–335.

Rawson, E.S., Freedson, P.S., Osganian, S.K., Matthews, C.E., Reed, G. and Ockene, I.S., 2003, Body mass index, but not physical activity, is associated with C-reactive protein. *Medicine and Science in Sports and Exercise*, **35**, 1160–1166.

Reilly, M.P. and Rader, D.J., 2003, The metabolic syndrome: More than the sum of its parts? *Circulation*, **108**, 1546–1551.

Reuben, D.B., Judd-Hamilton, L., Harris, T.B. and Seeman, T.E., 2003, The associations between physical activity and inflammatory markers in high-functioning older persons: MacArthur Studies of Successful Aging. *Journal of American Geriatric Society*, **51**, 1125–1130.

Rexrode, K.M., Pradhan, A., Manson, J.E., Buring, J.E. and Ridker, P.M., 2003, Relationship of total and abdominal adiposity with CRP and IL-6 in women. *Annals of Epidemiology*, **13**, 1–9.

Ridker, P.M., 2003, High-sensitivity C-reactive protein inflammation, and cardiovascular risk: From concept to clinical practice to clinical benefit. *American Heart Journal*, S19–S26.

Robinson, L.E. and Graham, T.E., 2004, Metabolic syndrome, a cardiovascular disease risk factor: Role of adipocytokines and impact of diet and physical activity. *Canadian Journal of Applied Physiology*, **29**(6), 808–892.

Ruderman, N., Chisholm, D., Pi-Sunyer, X. and Schneider, S., 1998, The metabolically obese, normal-weight individual revisited. *Diabetes*, **47**, 699–713.

Rutter, M.K., Meigs, J.B., Sullivan, L.M., D'Agostino, R.B. and Wilson, P.W.F., 2004, C-reactive protein, the metabolic syndrome, and prediction of cardiovascular events in the Framingham Offspring Study. *Circulation*, **110**, 380–385.

Saijo, Y., Kiyota, N., Kawasaki, Y., Kashimura, J. and Fukuda, M., 2004, Relationship between C-reactive protein and visceral adipose tissue in healthy Japanese subjects. *Diabetes, Obesity and Metabolism*, **6**, 249–258.

Seshadri, P., Iqbal, N., Stern, L., Williams, M., Chicano, K.L. and Daily, D.A., 2004, A randomized study comparing the effects of a low-carbohydrate diet and a conventional diet on lipoprotein subfractions and C-reactive protein levels in patients with severe obesity. *American Journal of Medicine*, **117**, 398–405.

Shen, B.J., Todaro, J.F., Niaura, R., Mc affery, J.M., Zhang, J. Spiro, A. and Ward, K.D., 2003, Are metabolic risk factors one unified syndrome? Modeling the structure of the metabolic syndrome X. *American Journal of Epidemiology*, **157**(8), 701–711.

Sites, C.K., Toth, M.J., Cushman, M., L'Hommedieu, G.D., Tchernof, A. and Tracy, R.P., 2002, Menopause-related differences in inflammation markers and their relationship to body fat distribution and insulin-stimulated glucose disposal. *Fertility and Sterility*, **77**, 128–135.

Slabbert, S., 2004, The relationship between traditional cardiovascular risk factors, body composition and C-reactive protein among 19–60 year old black women. (North-West University: Potchefstroom) 102 p.

Sobngwi, E., Mbanya, J-CN., Kengne, A.P., Fezeu, L., Minkoulou, E.M., Aspray, T.J. and Alberti, K.G.M.M., 2002, Physical activity and its relationship with obesity, hypertension and diabetes in urban and rural Cameroon. *International Journal of Obesity*, **26**, 1009–1016.

Taaffe, D.R., Harris, T.B., Ferrucci, L., Rowe, J. and Seeman, T.E., 2000, Cross-sectional and prospective relationships of Interleukin-6 and C-reactive protein with physical performance in elderly persons: MacArthur studies of successful aging. *Journal of Gerontology*, **55A**(12), M709–M715.

Tchernof, A., Nolan, A., Sites, C.K., Ades, P.A. and Poehlman, E.T., 2002, Weight loss reduces C-reactive protein levels in obese postmenopausal women. *Circulation*, **105**, 564–569.

Trayhurn, P. and Wood, I.S., 2004, Adipokines: Inflammation and the pleiotropic role white adipose tissue. *British Journal of Nutrition*, **92**, 347–355.

Vanhala, M.J., Vanhala, P.T., Keinanen-Keiukaanniemi, S.M., Kumpusalo, E.A. and Takala, M.J.K., 1999, Relative weight gain and obesity as a child predict metabolic syndrome as an adult. *International Journal of Obesity*, **23**, 656–659.

Verdaet, D., Dendale, P., De Bacquer, D., Delanghe, J., Block, P. and Backer, G. 2004. Association between leisure time physical activity and markers of chronic inflammation related to coronary heart disease. *Atherosclerosis*, **176**:303–310.

Vikram, N.K., Misra, A., Dwivedi, M., Sharma, R., Pandey, R.M., Luthra, K., Chatterjee, A., Dhingra, V., Jailkhani, B.L., Talwar, K.K. and Guleria, R., 2003, Correlations of C-reactive protein levels with anthropometric profile, percentage of body fat and lipids in healthy adolescents and young adults in urban North India. *Atherosclerosis*, **168**, 305–313.

Vozarova, B., Weyer, C., Hanson, K., Tataranni, P.A., Bogardus, C. and Pratley, R.E., 2001, Circulating interleukin-6 in relation to adiposity, insulin action, and insulin secretion. *Obesity research*, **9**(7), 414–417.

Wang, T.J., Larson, M.G., Levy, D., Benjamin, E.J., Kupka, M.J., Manning, W.J., Clouse, M.E., D'Agostino, R.B., Wilson, P.W.F. and O'Donnell, C.J., 2002, C-Reactive protein is associated with subclinical epicardial coronary calcification in men and women. *Circulation*, **106**, 1189–1191.

Wannamethee, S.H., Lowe, G.D.O., Whincup, P.H., Rumley, A., Walker, M. and Lennon, L., 2002, Physical activity and hemostatic and inflammatory variables in elderly men. *Circulation*, **105**, 1785–1790.

Weisberg, S.P., McCann, D., Desai, M., Rosenbaum, M., Leibel, R.L. and Ferrante, A.W., 2003, Obesity is associated with macrophage accumulation in adipose tissue. *Clinical Investigation*, **112**, 1796–1808.

Weyer, C., Yudkin, J.S., Stehouwer, C.D.A., Schalkwijk, C.G., Pratley, R.E. and Tataranni, P.A., 2002, Humoral markers of inflammation and endothelial dysfunction in relation to adiposity and in vivo insulin action in Pima Indians. *Atherosclerosis*, **161**, 233–242.

Wong, S.L., Janssen, I. and Ross, R., 2003, Abdominal adipose tissue: Distribution and metabolic risk. *Sports Medicine*, **33**(10), 709–726.

World Health Organization (WHO), 1998, *The World Health Report 1998*. Life in the 21st century: A vision for all (Geneva : Switzerland), 241 p.

World Health Organization (WHO), 2000, *Obesity: Preventing and Managing the Global Epidemic Report of a WHO Consultation of Obesity* (Geneva: Switzerland.) World Health Organization, 253p.

World Health Organization (WHO), 2003, Obesity and overweight. [Web:] http://www.who.int/dietphysicalactivity/publications/facts/obesity/en/ [Date of access: 19 March 2005].

Xydakis, A.M., Case, C.C., Jones, P.H., Hoogenveen, R.C. Liu, M.Y. and Smith, E.O., 2004, Adiponectin, inflammation, and the expression of the metabolic syndrome in obese individuals: The impact of rapid weight loss through caloric restriction. *Journal of Clinical Endocrinology and Metabolism*, **89**, 2697–2703.

Yeater, R., 2000, Obesity, metabolic syndrome and physical activity. *Quest*, **52**, 351–357.

Yudkin, J.S., Juhan-Vague, I., Hawe, E., Humphries, S.E., Di Minno, G., Margaglione, M., Tremoli, E., Kooistra, P.E., Morange, P.E., Lundman, V., Mohamed-Ali, V. and Hamsten, A., 2004, Low-grade inflammation may play a role in the etiology of the metabolic syndrome in patients with coronary heart disease: The HIFMECH study. *Metabolism*, **53**(7), 852–857.

Yudkin, J.S., Stehouwer, C.D.A., Emeis, J.J. and Coppack, S.W., 1999, C-reactive protein in healthy subjects: Associations with obesity, insulin resistance, and endothelial dysfunction. A potential role for cytokines originating from adipose tissue? *Arteriosclerosis, Thrombosis and Vascular Biology*, **19**, 972–978.

Ziccardi, P., Nappo, F., Giugliano, G., Esposito, K., Marfella, R and Cioffi, M., 2002, Reduction of inflammatory cytokine concentrations and improvement of endothelial functions in obese women after weight loss over one year. *Circulation*, **105**(7), 804–809.

CHAPTER FIFTEEN

Towards a generalised anthropometric language

T. Kupke and T. Olds

University of South Australia

1 THE PROLIFERATION OF ANTHROPOMETRIC PROTOCOLS

Over the past centuries many landmarks and dimensions have been created and adapted. Many researchers have developed their own measurement techniques and created a large number of dimensions applicable to their own disciplines and interests. This has led to an increasing variety of specialised anthropometric protocols spread throughout the world.

Unfortunately, many scientific disciplines have not been able to agree on a standardisation of anthropometric techniques. As the scientific community increases and branches out into more fields of interest, the number of protocols increases also. Ergonomics (human factors/engineering anthropometry), sport science, health and medical sciences, nutrition, forensics, biomechanics, psychology and anthropology are some of the disciplines using anthropometry as a basis for their research. The clothing industry and military research are two other major areas that use anthropometry and these are the driving forces for the technology advancements associated with anthropometry.

A few protocols use similar dimensions and landmarks, whereas others are quite different. Some protocols measure on the left side of the body, some on the right, some on both, and some do not specify which side of the body to measure. Some of the current major protocols include the International Society for the Advancement of Kinanthropometry (ISAK), the Civilian American and European Surface Anthropometry Resource (CAESAR), the US Army Anthropometric Survey 1987–1988 (ANSUR), the International Organization for Standardization (ISO) 8559, and the International Biological Programme (IBP).

This study aims to make researchers aware of the changing nature of anthropometric protocols around the world and the disconnection between its users. It aims to highlight the problems involved with this, and suggests a solution that may help all anthropometric researchers understand each other better. It creates a generalised anthropometric language (GAL) that allows all dimensions to be described in a systematic way, by using a three-dimensional logic.

1.1 The genealogy of anthropometric protocols: drift, divergence and convergence

Paul Broca (1824–1880), the founder of the "École d'Anthropologie" in Paris in 1859, helped to increase the interest in anthropometry around this time. Until the Franco-Prussian war of 1870, the Broca system was used almost universally (Roebuck *et al.*, 1975; Hrdlićka, 1920). After this war there became a "French school" protocol and a "German school" protocol. About 30 years later there were efforts to converge the protocols into an international anthropometric protocol. This led to the development of an International Agreement on Anthropometry at conferences in 1906 in Monaco (Papillault, 1906) and 1912 at Geneva (Duckworth, 1912), with the International Congress of Prehistoric Anthropology and Archeology. The 1906 conference produced a protocol for measuring the skull (craniometry) and living head and facial features (cephalometry), while the 1912 conference developed technique for measuring the shape (somatometry) and bones of the body (osteometry). These publications were aimed at becoming the internationally recognised protocol for some time.

However, soon after, in 1914, Rudolf Martin wrote his book "Lehrbuch der Anthropologie", which became the prominent handbook for several decades (Kroemer *et al.*, 1994). Martin followed the 1906 and 1912 agreements for most of his work (Roebuck *et al.*, 1975). Martin subsequently produced a three-volume revised edition in 1928, and then a two-volume revised edition in 1957 with Karl Saller. Aleś Hrdlićka published the 1906 and 1912 agreements again in 1920 (Hrdlićka, 1920), and also included his own handbook that followed the agreements with slight modifications and suggestions.

The Martin handbooks have become the basis of many current protocols with many modifications included over the years. Many further anthropometry handbooks have been published since, and whilst still using the underlying techniques and definitions used in the early 1900's they have subtly drifted apart from one another.

The genealogy of anthropometric protocols displays a complex combination of definitions from many sources. However, with the 1906 and 1912 agreements as a basis, followed by Martin in 1914, many new protocols have stemmed from these works, with divergence over the years. A chronological outline of this genealogy is shown in Figure 15.1. Discipline areas are shown on the most recent anthropometric protocols. These indicate the area in which the protocol is most commonly used.

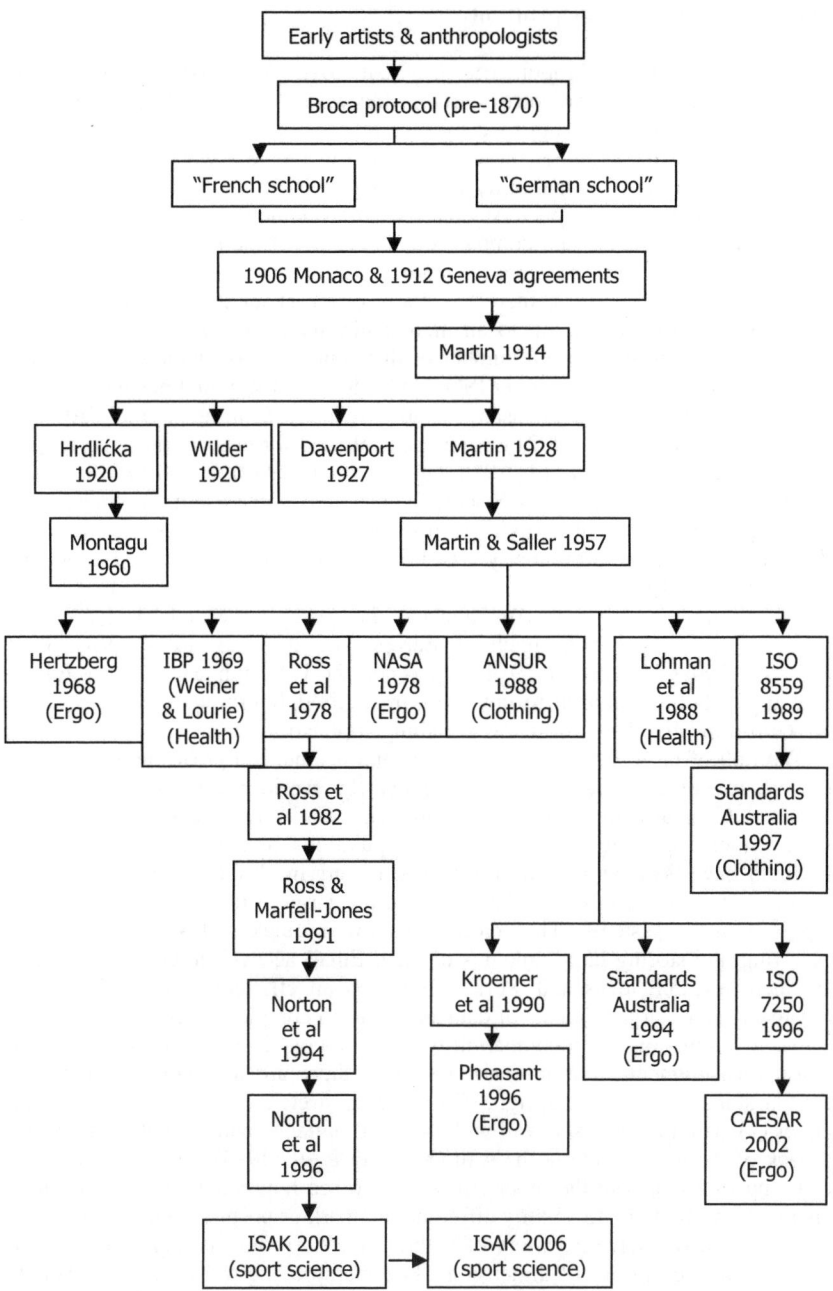

Figure 15.1 The genealogy of anthropometric protocols. (Ergo = Ergonomics, Clothing = Clothing industry.)

1.2 Problems with multiple protocols

The proliferation of anthropometric protocols over the past 150 years has disconnected many anthropometrists around the world. With the large number of anthropometric studies accomplished by researchers around the world it would be advantageous to bring together these studies and to create a worldwide database. Unfortunately, without compatibility of the protocols, this is a lost opportunity. There have been many attempts to standardise anthropometric protocols in the past, but it is difficult to collate and compare them because of the increasing number of different dimensions being measured in studies.

Even within disciplines, there are many different anthropometric protocols used around the world, leading to an incompatibility of databases. For example, currently in the U.S. there are a number of different protocols widely used. In the ergonomics area many follow either ISO 7250 (1996) or the guidelines published in Kroemer *et al.* (1990) and Pheasant (1996). In the U.S. military, the CAESAR protocol (Blackwell *et al.*, 2002) has been established for ergonomics whereas the ANSUR protocol (Clauser *et al.*, 1988, Gordon *et al.*, 1989) of 1987/1988 is used for clothing applications. In the US sport science area, some use the Lohman *et al.* (1988) techniques, or the ISAK protocol (International Society for the Advancement of Kinanthropometry, 2001), and others use specialised protocols from the 1960s and 1970s.

The clothing industry in Australia has the Australian Standards (AS 1344, 1997) to follow. However, much of Europe and the U.S. follow ISO 8559 (International Standard ISO 8559, 1989). Again the U.S. military does not follow any of these, but has chosen to create its own ANSUR protocol (Clauser *et al.*, 1988, Gordon *et al.*, 1989) for its own clothing applications.

With all of these different applications and protocols a great opportunity has been lost to combine data from throughout the world across many scientific disciplines. Comparisons of body size and shape could have been made between countries, races, sporting groups, diseased populations, and socio-economic status that would have become a great help to researchers in all these areas. More than ever before, we have access to quick data transfer, large storage of information and complex analysis systems. The internet is now a quick and simple means of transferring and storing large amounts of data. Email access allows researchers to share their research ideas and work together in an efficient manner. Computer software can store large amounts of data and perform complex analyses.

An example of the incompatibility of protocols is found in some of the simplest measurements. For measuring stature, there are four common methods used: (1) standing naturally upright, (2) standing stretch stature, (3) lying on the back, or (4) standing against a wall with the back flattened and buttocks, shoulders, and back of the head touching the wall (Kroemer *et al.*, 1994). Each of these may vary by up to 2 cm from the others. It is easy to see how many other dimensions cannot be compared due to slightly different landmark or postural definitions.

While we still have a variety of anthropometric protocols there will also be confusion over definitions of dimensions. For example, the waist-hip ratio (WHR) is often used in public health. However there are many differing definitions of where the waist and hip girths for this ratio are to be measured. Also some protocols do not define their landmarks as precisely as others do. For example, ISO 8559 often refers to one end of a dimension as the knee, not specifying where on the knee.

The WHR is often used, and widely recommended in clinical practice as a tool for determining fat distribution (Alexander and Dugdale, 1990). This leads to recommendations about the associated risks of heart disease and death amongst the general public. However, the actual sites of measuring the waist and hip circumferences are rarely defined or standardised. Alexander and Dugdale (1990) used two waist and two hip circumference definitions amongst a group of obese females, creating four unique WHR definitions. They discovered that the mean WHR of the group ranged from 0.76 ± 0.04 (mean ± standard deviation) to 0.95 ± 0.03, depending on which WHR definition was used. Depending on the definition, 23% to 100% of the females were classified in the "at risk" category. This shows that a lack of precision in definitions can make a practical difference to the manner in which the professional health community operates.

A further example of confusion in the sport science area is the dimension of foot length as defined by the ISAK protocol, which has been interpreted in two ways. A straight-line point to point measure (from the Akropodion [anterior tip of the longest toe] to the Pternion [posterior point on the rear of the foot]) and a perpendicular distance measure (the distance between the vertical planes of the Akropodion and the Pternion), have both often been used. This is just one example of an internationally known protocol being misunderstood by its followers. A recent revision of the ISAK protocol (Marfell-Jones *et al.*, 2006) explains the new definition for foot length will become "the perpendicular distance between the coronal planes of the Pternion and Akropodion".

Currently there is no syntax available for two-dimensional (2D) and three-dimensional (3D) anthropometry. Measures such as surface areas, cross-sectional areas and volumes are becoming more common and more accessible as the use of 3D scanning technology increases. Producing universal definitions and consistent methods of describing the techniques used for measuring these 2D and 3D dimensions will be beneficial before further work is completed in the area.

There are advantages and disadvantages to standardising anthropometric protocols. A standardisation of protocols would create more opportunity for sharing and comparing anthropometric databases. Anthropometry is similar to every other science, which should adhere to a protocol in order to ensure that others in the same area are performing in the same manner (Norton and Olds, 1996; Hrdlićka, 1920) suggested that one of the aims of those interested in anthropometry should be to develop a general unification of instruments and techniques, as much as is practical. The IBP protocol (Weiner and Lourie, 1969) was designed to help with gaining data that would be comparable across all populations in all environments.

However, Weiner and Lourie (1969) also encouraged the researchers who use their work, not to limit themselves to their protocol, but to also extend on it with their own ideas and dimensions. Because there are so many scientific disciplines now using anthropometry there are a greater number of dimensions of interest to the researchers, making standardisation more difficult. Gordon and Friedl (1994) suggested that many classical anthropometric dimensions were not applicable to military use, and therefore combining measuring techniques with others was useless. Hrdlićka (1920) suggested that anthropometry would never be a universal system, because the dimensions that one researcher finds useful and interesting may be of no use to another. Many other researchers may take this view and continue to create their own techniques, but sometimes not explicitly defining them in their published work.

Researchers need to become aware of the seeming carelessness that has gradually developed in performing and reporting of anthropometric techniques. To rectify this, firstly, researchers need to become aware, if they are not already, of the variety of anthropometric protocols, and techniques within them, which are used throughout the world; and, secondly, researchers need to report their anthropometric techniques in detail using anatomical terms, or refer to a specific publication that includes the anthropometric protocol chosen. Alternatively, and vastly different from the other two, a GAL could be developed to either combine protocols or at least bring them all together under a single language that all researchers could understand. It is this option that this paper will explore.

2 TOWARDS A GENERALISED ANTHROPOMETRIC LANGUAGE – DIMENSIONAL SYNTAX

An anthropometric dimensional syntax is a system combining a variety of rules that help define all types of dimensions in terms of landmarks. Landmarks are defined by using a description of anatomical terms and sites. The syntax sets out the manner in which any number of landmarks can be combined to create a dimension. The combination of a series of dimensions makes up an anthropometric protocol.

There is a need to make anthropometric protocols comparable to each other. A GAL would help establish these links between protocols and therefore also between studies. A well-defined GAL with syntactic rules would help instruct researchers to refine and explicitly record their techniques. It would also decrease the incompatibility and confusion that currently exists between researchers using different anthropometric protocols. There is no limit to the number of landmarks available on the human body, especially with the increased use of 3D whole-body scanning producing XYZ coordinates. There is also an endless list of possible dimensions to be measured on the body. For this reason an expandable, flexible syntax would need to be designed.

A GAL would contain a "superset" of landmarks, a collection of landmarks from many protocols. The superset would be inclusive, rather than exclusive, and all the landmarks would be reviewed carefully to determine if multiple landmarks could be combined into one. Researchers could add to the superset by submitting their new landmark definitions to a central source for review. No doubt the superset would frequently change as researchers added their own landmark definitions. However, the landmark superset and dimensional syntax could be widely available around the world through an internet site, and changes could be accessed quickly without waiting for new printed publications. The superset would be available to all researchers who could draw upon useful subsets of landmarks for their own specific research goals. The GAL would help all researchers describe their protocols in a form that could be understood by other researchers.

A GAL would combine all anthropometric protocols into the same language by using the same syntax. Because of the diverse range of protocols available, a GAL would need to be flexible enough to include all dimensions and landmarks, including those that have less explicit definitions. A GAL that was totally inclusive would require:

(1) A "superset" of landmarks sufficient to cover those used across all the protocols (including those landmarks that are less explicit), and
(2) A "syntax" that describes how the dimensions can be defined with the use of landmarks.

A GAL should be flexible enough so that the anthropometric vocabulary can be expanded to include new landmarks in the future. It should also be a comprehensive language that has the potential to describe all body dimensions. It is important to note that a GAL is not just another anthropometric protocol, increasing the number of landmarks and dimensions. Rather, a GAL may help describe all major anthropometric protocols by using the same syntax. A GAL may also help to provide a framework for creating a common terminology across protocols, using a superset of landmarks from different protocols.

2.1 Three-dimensional logic

Most current anthropometric protocols are based on a one-dimensional system. They do not introduce the idea of using two or three dimensions for measurements of the human body. The GAL proposed in this paper (henceforth referred to as "the GAL"), is based on three-dimensional logic, currently used in many disciplines, but rarely defined in published anthropometric protocols. The dimensional syntax includes definitions of three-dimensional logic, describing:

- linear distances as the 3D Cartesian distances between landmarks
- contours and girths as the intersection of planes and the surface of a 3D body
- cross-sectional areas as the intersection of planes and a part of a 3D body
- surface areas as areas on a 3D body surface defined by planes or points
- volumes as parts of the 3D body defined by planes

Previous efforts to standardise anthropometric techniques (see Figure 15.1) have generally been exclusive, causing an incompatibility with many disciplines. They have often only included the landmarks and dimensions used by the majority of researchers in one discipline area, leaving other disciplines without some of their required definitions. This caused researchers to create and publish their own standards within the scientific community, hence, the multiple protocols available today. The GAL contains the landmark superset, which is inclusive rather than exclusive, and allows all researchers to draw upon any number of landmarks as a subset for their research. (See section three for more information on the landmark superset.)

The GAL is designed to allow for future analysis by computer software. The syntax outlined below in section 2.2 following, shows that all types of dimensions can now be described in a form that relies simply on a set of rules combining numbers and notations, rather than using descriptive words. It has especially been designed to be programmable in order to facilitate data storage, transfer and analysis.

2.2 Dimensional syntax

Throughout the GAL, the following typographic conventions have been used:

- The GAL is written in "Century Gothic" font, for example: distance
- A function is a means of defining dimensions. Functions are indicated by parentheses: distance() is a function. Other functions include: plane(), pdistance(), contour(), girth(), skinfold(), Xsection(), surface(), volume(), average(), max(), min().
- A function contains a number of variables within the parentheses. The three basic variables are landmarks (indicated by x_i, which designates the ith landmark), plane orientations (indicated by y), and posture (indicated by z).
- Each landmark (x) has been allocated a unique number within a landmark superset. The superset contains all possible landmarks combining the major anthropometric protocols. The landmark superset has the ability to expand, by allowing all researchers to submit their own definitions for the superset. Landmark definitions from different protocols that are found to have the same location have been allocated the same landmark number in the superset. (See section three for more information on the landmark superset.)
- A letter (R, L or M) follows each landmark number in subscript font, indicating whether the landmark is on the right (R), left (L), or midline (M) of the body.
- Plane orientations (y) are shown as relative to the plane of the standing surface (S) or relative to a plane perpendicular to the long axis of the limb or body segment (L) in the traditional anatomical standing posture.
- Posture (z) is expressed as anatomical standing posture (A), relaxed standing posture (B), standard scanning posture (C), relaxed seated posture (D), or upright seated posture (E). It is a compulsory variable, necessary in every function, and must always be the first variable stated.
- Variables inside angle brackets and separated by the word "or" represent alternatives. For example <x_1 or x_2 or x_3> would indicate that one of the variables x_1, x_2 or x_3 must be inserted at this point in the syntax.

2.2.1 Example

In the expression pdistance(A,51R,plane(75R,S0r0f)), the three variables are A (describing the posture as the anatomical standing posture), 51_R (describing landmark 51_R) and plane(75R,S0r0f) (describing a plane through landmark 75_R, with a plane orientation of S0r0f (describing a plane parallel to the standing surface (see plane orientation in section 2.2.2 below))). Variables are to be separated by commas. pdistance() is the main function and it refers to a perpendicular distance. In this example, there is a function within a function. Therefore posture A, landmark 75_R and plane orientation S0r0f are also variables within the plane function.

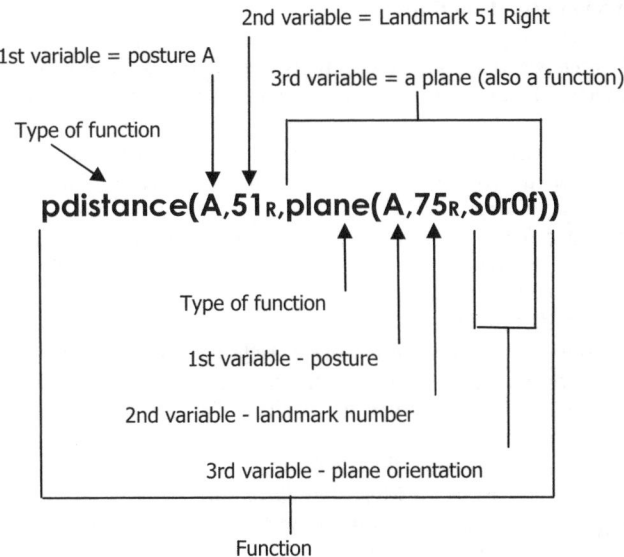

Figure 15.2 An example of a function showing the notation used for the dimensional syntax.

2.2.2 Plane orientation

Plane orientation is shown relative to a reference plane. This reference plane may be either the plane of the standing surface (S) or the plane perpendicular to the long axis of the limb or body segment (L) in the traditional anatomical position. To define a plane, two angles must be specified: the anti-clockwise angle the plane makes with the right-left axis (r) of the reference plane, and the anti-clockwise angle it makes with the front-back (f) axis of the reference plane.

To uniquely define a plane, the perspective that the measurer takes must be consistent. Whenever an angle is specified relative to the right-left axis (r), we define that the measurer must create the angle while viewing the front of the subject. Whenever an angle is specified relative to the front-back axis (f), we define that the measurer must create the angle while viewing the right hand side of the subject. Taking these perspectives ensures that the anti-clockwise angle made between the axes of the created plane and the reference plane defines a unique plane.

A plane orientation has the form: $y = <S$ or $L> \alpha r \beta f$, where S or L indicates the reference plane, α is the angle (in degrees) relative to the right-left axis (r) of the reference plane, and β is the angle (in degrees) relative to the front-back axis (f) of the reference plane.

For example, L0r0f would specify a plane exactly perpendicular to the long axis of the body or body segment with no deviation on either axis. S0r0f would specify a plane exactly parallel to the standing surface with no deviation on either axis.

2.3 Anthropometric functions

The GAL consists of a set of functions for defining dimensions. These define anthropometric dimensions such as breadths, girths, lengths, contours, surface areas, cross-sectional areas and volumes by reference to landmarks and plane orientations. Functions can be used within another function, and this can be expanded as much as necessary. Functions can also be combined with the use of standard arithmetic operators. For example, adding or subtracting distances can be done to account for the use of an anthropometry box during traditional physical measurements. The syntax for a girth function and a skinfold function are shown below, including an example of each.

Syntax: girth(z,x,y)

Description: This function defines a girth, measured in posture z, which is the intersection between a plane with orientation y and the body, starting and ending at the same landmark x. A girth is the surface distance around the body or body segment.

Example: The function girth(B,65$_M$,S0r0f), defines the girth circumference, at the level of landmark 65$_M$ (Mesosternale) along a line where a plane parallel to the standing surface intersects with the body. This represents the girth around the chest at the level of the Mesosternale landmark measured in the relaxed standing posture (B) (see Figure 15.3).

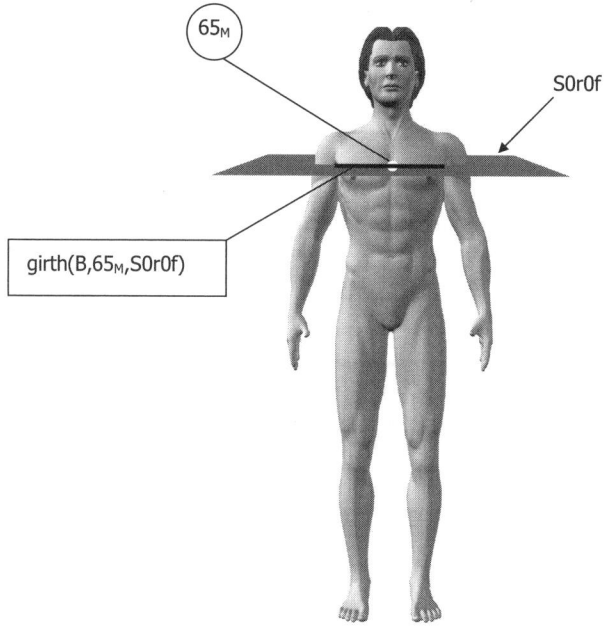

Figure 15.3 girth(B,65$_M$,S0r0f).

Syntax: skinfold(z,x,y)

Description: A skinfold is the linear depth of a double fold of skin and subcutaneous fat, taken at a certain landmark (x), along the path where a plane with orientation y intersects the body through that landmark. The posture is indicated by z.

Example: The function skinfold(B,73ᵣ,L45r0f), defines the skinfold taken at landmark 73ᵣ (Subscapular Skinfold Site Right), in a plane orientated at 45° to the long axis of the trunk. This represents the Subscapular skinfold measured in the relaxed standing posture (B). Note that the angle stated is referring to the skinfold, whereas the angle of the calipers is 90° to this (see Figure 15.4).

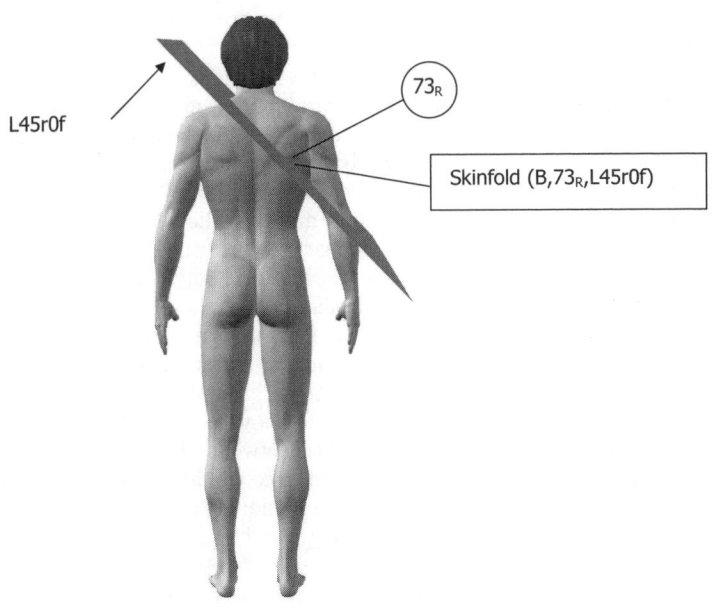

Figure 15.4 skinfold(B,73ᵣ,L45r0f).

3 TOWARDS A GENERALISED ANTHROPOMETRIC LANGUAGE – LANDMARK SUPERSET

If a GAL is to be understood by many researchers around the world, it must be flexible and comprehensive enough to include the majority of landmarks used in different scientific disciplines. A GAL requires a landmark superset, which includes all the necessary landmarks to help define dimensions within the GAL. The superset is a combination of landmarks from many protocols, which is refined to exclude those landmarks regarded as the same as each other in order to simplify the superset. As shown in section one, there is a diverse range of anthropometric protocols currently used around the world. In this study, the landmark superset has been collated from all the landmarks from some of the major current anthropometric protocols.

To create a landmark superset a number of steps were taken. Firstly, the current major anthropometric protocols were identified and then five were chosen for inclusion in this landmark superset. Secondly, every landmark from all five of these protocols was theoretically compared to each of the others according to the description of anatomical terms in their written definitions. This included a total of 359 different landmarks (including explicit and inexplicit landmarks) relating to 383 different dimensions. All 359 landmarks were put through a theoretical comparison process and those landmarks that showed equivalencies and similarities to others were noted. Thirdly, some of the landmarks that were found to be theoretically similar were physically landmarked on a sample of 20 subjects to see if they could be regarded as equivalent landmarks. Differences between the similar landmarks were measured on the subjects. A sample of only six physical comparisons was used to test the theory and to show its capability for further use. Fourthly, statistical comparisons were made between the physically compared landmarks to analyse whether two similar landmarks could be regarded as the same one. Finally, the landmarks in the superset were uniquely numbered and named, and those landmarks showing equivalencies were given the same number and name.

The five major anthropometric protocols chosen for inclusion in this landmark superset were: CAESAR (Blackwell, 2002; Robinette, 2002), ISAK (ISAK, 2001), ANSUR (Clauser *et al.*, 1988; Gordon *et al.*, 1989), ISO 8559 (International Standard ISO 8559, 1989), and IBP (Weiner and Lourie, 1969). This choice was based on the advice of various anthropometry experts who were asked to name what they saw as the major anthropometric protocols.

3.1 Theoretical comparisons

A theoretical comparison of all the landmarks from the five anthropometric protocols was performed by studying the description of the anatomical terms in their written definitions. Firstly, all of the explicit landmark definitions were noted from each protocol. Secondly, all of the dimension definitions were studied to find any end points or reference points that had not been explicitly defined as landmarks in the protocol. Thirdly, every landmark and dimension was numbered, and then cross-tabulated to show the appropriate landmarks or dimensions associated with each one.

With all 359 of the possible landmarks found, all of the landmarks were compared to others to find any that could be regarded as equivalent or similar due to their written definition. For example, CAESAR's Outer Corner of Eye (84) landmark (defined as "The outer corner of the eye") was compared with ANSUR's Ectocanthus (104) landmark (defined as "The outside corner of the right eye formed by the meeting of the upper and lower eyelids"). Study of their definitions resulted in the two landmarks being regarded as equivalent landmarks. Sometimes the definitions of two landmarks differed enough to make them non-equivalent even though they were similar. For example, ISAK's Acromiale (1) landmark (defined as "The point on the superior part of the acromion border in line with the most lateral aspect") was found to be similar but not equivalent to IBP's Acromion (4) landmark (defined as "The inferior edge of the most external border of the acromion process").

Once all the theoretical comparisons were completed, those landmarks that were equivalencies were combined to represent a single landmark in the superset. All of the landmarks in the superset were then given a unique landmark number and name, with the name usually stemming from the original landmark name in its protocol. To further refine the landmark superset, physical comparisons of landmarks were applied.

The collated list of landmarks began with 359 landmarks from the five protocols. After completing the theoretical comparisons, the superset had been refined to 162 unique landmarks.

The landmark superset was then complete and ready to be used in the GAL by incorporating the appropriate landmarks when defining dimensions. This landmark superset can be obtained from the University of South Australia library in "Kupke, T., 2005, Towards a generalised anthropometric language. Unpublished Honours Thesis. Adelaide, SA: University of South Australia", or by emailing tim.kupke@unisa.edu.au.

3.2 Physical comparisons

Some of the landmarks that were shown to be similar in the theoretical comparison tables were chosen for physical comparison on 20 subjects. Six landmark areas were chosen for physical comparison including the Acromion, Cervicale, Olecranon, Anterior Superior Iliac Spine (ASIS), Iliocristale, and Trochanterion areas. These landmark areas were chosen because of the regular use of these landmarks in anthropometry. Within each landmark area, two landmarks, each from different protocols, were chosen for physical comparison. The two landmarks chosen were noted as similar but not equivalent during the theoretical comparisons. Only six landmark areas were chosen so that each subject could have the procedure completed within 45 minutes.

Twenty subjects were recruited for the physical comparisons of landmarks. Three ISAK-accredited Level 2 anthropometrists were used for this trial. The subjects were landmarked on the 12 landmarks with repeated measures on each definition.

The subjects were landmarked by using a special marker known as a "UV light" or "Black light" pen. The anthropometrists landmarked each subject four times in the one landmark area, including repeated measures on each definition. Each time they replicated the landmark they would not know where previous marks were placed, because the marks produced by the pens were not visible in normal lighting. Once the landmarking was completed on the subject, the lights were dimmed and a UV light was used to display the marks applied. While visible, the marks were then reapplied as a single fine dot with pen. The lights were then brightened allowing the ink pen marks to be visible, and the distances between them to be measured.

The distances between each of the four marks in the one landmark area were measured using a Lufkin girth tape, and recorded in mm. The four landmarks create a total of six distances – two intra-definition distances, between the repeated measures of each of the landmark definitions, and four inter-definition distances, between landmarks of differing definitions.

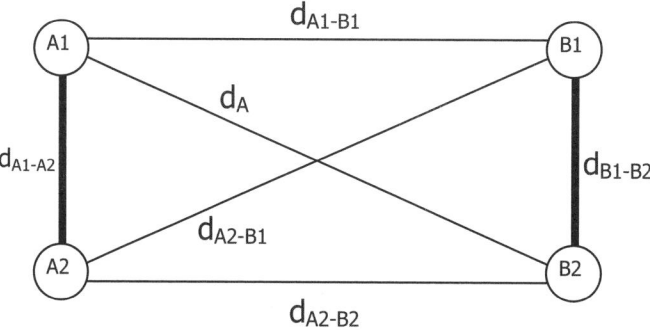

Figure 15.5 An example of a landmarked area. It includes four landmarks – A1, A2, B1, and B2. The thicker lines indicate the two intra-definition distances, and the thinner lines indicate the four inter-definition distances.

Figure 15.5 shows one landmark area. This includes repeated measures on two separate definitions. Definition A is marked twice (shown by A1 and A2), and definition B is marked twice (shown by B1 and B2). The two intra-definition distances are those that are between two marked sites of the same definition. In Figure 15.5, the intra-definition distances are shown as d_{A1-A2} and d_{B1-B2}, and are shown as the thicker lines. The four inter-definition distances are those that are between two marked sites of different definitions. The inter-definition distances are shown as d_{A1-B1}, d_{A1-B2}, d_{A2-B1}, and d_{A2-B2}, and are shown as the thinner lines. This process is applied to every landmark area for every subject and with each anthropometrist.

Statistical analysis was based on a comparison between the mean intra-definition distances and the mean inter-definition distances for each of the landmark areas. A mixed-design ANOVA test was used for the comparisons, with anthropometrist (A, B, or C) as the grouping factor, and intra-definition distances vs inter-definition distances as the repeated measure on each subject. The alpha level was set at 0.05. This test was used to discover whether the two physically compared landmarks in each landmark area were significantly different from each other. A Bonferroni correction was applied to results of the ANOVA tests. This was applied to counter the "alpha slippage".

The results showed that the mean inter-definition distances were significantly greater than the mean intra-definition distances for every landmark area (Acromion, Cervicale, Olecranon, ASIS, Iliocristale, and Trochanterion), i.e. the two compared definitions were not close enough to be regarded as the same landmark. In each case the difference between the two definitions was significantly larger than the measurement error involved in landmarking the same definition twice. Therefore within each landmark area, the definitions were regarded as similar but not equivalent, and they had to remain as separate landmarks in the landmark superset.

It must be noted that the number of subjects and anthropometrists used in this study was limited by the availability of resources. If larger numbers of subjects and anthropometrists had been used, we could have been more confident with the conclusion of the physical comparisons. Furthermore, this could have affected the decision as to whether two landmarks were regarded as "equivalent" or "similar but not equivalent".

4 FUTURE USE OF THE GAL

There is a risk that more anthropometric protocols will be formed in the future, each mutually exclusive, creating a larger web of protocols available across different scientific disciplines. The need for a GAL is clear through the incompatibility, confusion, and imprecision that exists with the current multitude of protocols. While 3D whole-body scanning utilises two- and three-dimensional measurements the GAL needs to consider these measures.

The GAL created in this study has shown that through the use of an inclusive landmark superset and a well defined dimensional syntax, any type of dimension on the body can be described in a form suitable for all researchers to understand. The mechanisms of producing a landmark superset outlined in this study should be imitated by an international organisation associated with anthropometry. A group knowledgeable in the theoretical aspects of anthropometry could perform the theoretical comparisons. The physical comparisons could then also be performed by a group of accredited anthropometrists, each individual with substantial practical experience. The organisation would need to communicate the GAL to other organisations involved in anthropometry, and keep all researchers regularly updated to changes in the superset through an internet based site.

Finally, the GAL used in this study is ready to be programmed into computer software. A program could be created to help researchers translate their current protocol definitions into the GAL. The interface would include lists of definitions from all anthropometric protocols, giving options to translate landmarks and dimensions into the appropriate superset landmarks and dimensional syntax.

5 REFERENCES

Alexander, H. and Dugdale, A.E., 1990, Which waist-hip ratio? [letter]. *The Medical Journal of Australia,* **153**, 367–368.
Blackwell, S., Robinette, K.M., Boehmer, M., Fleming, S., Kelly, S., Brill, T., Hoeferlin, D., Burnsides, D. and Daanen, H., 2002, *Civilian American and European Surface Anthropometry Resource (CAESAR), Final Report, Volume II: Descriptions* (AFRL-HE-WP-TR-2002-0173, Air Force Research Laboratory, Human Effectiveness Directorate, Crew System Interface Division, 2255 H Street, Wright-Patterson AFB OH 45433-7022 and Society of Automotive Engineers International, 400 Commonwealth Drive, Warrendale, PA: United States Air Force Research Laboratory).
Clauser, C., Tebbets, I., Bradtmiller, B., McConville, J., and Gordon, C.C., 1988, *Measurer's Handbook: US Army Anthropometric Survey 1987–1988* (Natick, MA: United States Army Natick Research, Development and Engineering Center).
Davenport, C.B., 1927, *Guide to Physical Anthropometry and Anthroposcopy.* (Eugenics Research Association Handbook Series). (New York: Waverly Press).
Duckworth, W.L.H., 1912, *The International Agreement for the Unification of Anthropometric Measurements to be Made on the Living Subject* (The Anthropological Laboratory of the University, New Museums, Cambridge. Also found in Hrdlička, 1920).
Gordon, C.C. and Friedl, K.E., 1994, Anthropometry in the US armed forces. In, *Anthropometry: The Individual and the Population*, edited by S.J. Ulijaszek, and C.G.N. Mascie-Taylor (New York: Cambridge University Press), pp. 178–210.
Gordon, C.C., Bradtmiller, B., Churchill, T., Clauser, C.E., McConville, J.T., Tebbetts, I.O. and Walker, R.A., 1989, *1988 Anthropometric Survey of US Army Personnel: Methods and Summary Statistics (Technical Report NATICK/TR-89/044)* (Natick, MA: US Army Natick Research, Development and Engineering Center).
Hertzberg, H.T.E., 1968, The conference on standardization of anthropometric techniques and terminology. *American Journal of Physical Anthropology,* **28**(1), 1–16.
Hrdlička, A., 1920, *Anthropometry* (Philadelphia, PA: The Wistar Institute of Anatomy and Biology).
International Society for the Advancement of Kinanthropometry (ISAK), 2001. *International Standards for Anthropometric Assessment* (Underdale, SA, Australia).
International Standard ISO 7250, 1996, *Basic Human Body Measurements for Technological Design* (Geneva: International Organization for Standardization).
International Standard ISO 8559, 1989, *Garment Construction and Anthropometric Surveys – Body Dimensions* (Geneva: International Organization for Standardization).
Kroemer, K.H.E., Kroemer, H.B. and Kroemer-Elbert, K.E., 1994, *Ergonomics: How to Design for Ease and Efficiency* (Englewood Cliffs, NJ: Prentice Hall).
Kroemer, Kroemer, and Kroemer-Elbert, 1990, *Engineering Physiology: Bases of Human Factors/Ergonomics*, 2nd ed. (New York: Van Nostrand Reinhold).

Lohman, T.G., Roche, A.F. and Martorell, R., 1988, *Anthropometric Standardization Reference Manual* (Champaign, IL: Human Kinetics Books).

Marfell-Jones, M.J., Olds, T., Stewart, A.D. and Carter, L., 2006, International Standards for Anthropometric Assessment, International Society for the Advancement of Kinanthropometry (ISAK) (Potchefstroom, South Africa).

Papillault, G., 1906, The international agreement for the unification of craniometric and cephalometric measurements. *L'Anthropologie, 17*, 559–572. (Also found in Hrdlićka, 1920).

Pheasant, S., 1996, *Bodyspace: Anthropometry, Ergonomics and the Design of Work*, 2nd ed. (London: Taylor & Francis).

Robinette, K.M., Blackwell, S., Daanen, H., Boehmer, M., Fleming, S., Brill, T., Hoeferlin, D. and Burnsides, D., 2002, *Civilian American and European Surface Anthropometry Resource (CAESAR), Final Report, Volume I: Summary* (AFRL-HE-WP-TR-2002-0169, Air Force Research Laboratory, Human Effectiveness Directorate, Crew System Interface Division, 2255 H Street, Wright-Patterson AFB OH 45433-7022 and Society of Automotive Engineers International, 400 Commonwealth Drive, Warrendale, PA: United States Air Force Research Laboratory).

Roebuck, J.A., Kroemer, K.H.E. and Thomson, W.G., 1975, *Engineering Anthropometry Methods* (Toronto: John Wiley & Sons).

Martin, R., 1914, *Lehrbuch der Anthropologie in Systematischer Darstellung* (Jena: Gustav Fischer).

Martin, R., 1928, *Lehrbuch der Anthropologie* (3 vols.) (Jena: Gustav Fischer).

Martin, R. and Saller, K., 1957, *Lehrbuch der Anthropologie* (2 vols.) (Stuttgart: Gustav Fischer).

Montagu, M.F.A., 1960, *A Handbook of Anthropometry* (Springfield, IL: Charles C. Thomas).

National Aeronautics and Space Administration (NASA), 1978, *Anthropometric Source Book, Vol.1: Anthropometry for Designers, Vol. 2: A Handbook of Anthropometric Data, Vol 3: Annotated Bibliography* (NASA Ref. Pub. 1024) (Houston, TX: NASA).

Norton, K.I. and Olds, T.S., 1996, *Anthropometrica* (Sydney, Australia: UNSW Press).

Norton, K.I., Whittingham, N., Carter, L., Kerr, D. and Gore, C., 1994, Measurement techniques in anthropometry. In *Anthropometry and Anthropometric Profiling*, edited by Norton, K.I. and Olds, T.S. (Sydney, Australia: Nolds Sports Scientific), pp. 1–32.

Norton, K.I., Whittingham, N., Carter, L., Kerr, D., Gore, C. and Marfell-Jones, M., 1996, Measurement techniques in anthropometry. In *Anthropometrica*, edited by Norton, K.I. and Olds, T.S. (Sydney, Australia: UNSW Press), pp. 25–75.

Ross, W.D. and Marfell-Jones, M.J., 1991, Kinanthropometry. In *Physiological Testing of the High Performance Athlete*, edited by MacDougall, J.D., Wenger, H.A. and Green, H.J., 2nd ed. (Champaign, IL: Human Kinetics Books), pp. 223–308.

Ross, W.D., Marfell-Jones, M.J. and Stirling, D.R., 1982, Prospects in Kinanthropometry. In *Sport Sciences*, edited by Jackson, J.J. and Wenger, H.A. (Victoria, Canada: School of Physical Education, University of Victoria), pp. 134–150.

Ross, W.D., Hebbelinck, M., Brown, S.R., and Faulkner, R.A., 1978, Kinanthropometric landmarks and terminology. In *Fitness Assessment*, edited by Shepard, R.J. and Lavallee, H. (Springfield, IL: Charles C. Thomas), pp. 44–50.

Standards Australia SAA HB59, 1994, *Ergonomics – the Human Factor: A Practical Approach to Work Systems Design* (Sydney, NSW: Standards Australia (Standards Association of Australia)).

Standards Australia, AS 1344., 1997, *Size Coding Scheme for Women's Clothing - Underwear, Outerwear and Foundation Garments* (Sydney: Standards Australia (Standards Association of Australia)).

Weiner, J.S. and Lourie, J.A., 1969, *Human Biology: A Guide to Field Methods (IBP Handbook No. 9)* (Oxford: Blackwell).

Wilder, H.H., 1920, *A Laboratory Manual of Anthropometry* (Philadelphia, PA: Blakiston's Son and Company).

CHAPTER SIXTEEN

Physique relationships in body dissatisfaction

A.D. Stewart[1], A.M. Johnstone[2], K. Giles[2] and P.J. Benson[3]

[1]School of Health Sciences, The Robert Gordon University, Aberdeen, UK
[2]Division of Obesity and Metabolic Health, Rowett Research Institute, Aberdeen, UK
[3]School of Psychology, University of Aberdeen, Aberdeen, UK

1 INTRODUCTION

Body image has been defined as "an evaluation of body size, weight or other aspect of the body that determines physical appearance" (Thompson, 1990). It embraces the separate *perception* of these determinants in a visuo-spatial sense, and *satisfaction* of the perception. As a determinant of self-esteem, body image includes perceptual, affective and cognitive components which rely in part on the construction of a dual model, which represents oneself and others. The complex interaction of these components relies in part on 'objective anthropometry' (Kay, 1996), – in other words, a reliable measurement of the phenotype, against which other influences such as environment, experience, gender and personality are brought to bear. The representations of self and others are counterpoised, and a comparison 'set point' is the theoretical fulcrum of the combined influences, which determine the response (Kay, 1996).

A large array of instruments to assess body image has been developed over time. Different methods have different strengths and weaknesses, and occupy different positions along a spectrum of complexity and convenience. Questionnaires can address individual shape concerns, but cannot objectively assess perception. Visual analogue scales can address perception of dimensions, but are limited in capturing body shape. Figural stimuli in the form of photographs, line drawings or silhouettes have been constructed to depict a range of body size or shape from which the subjects selects a representative or ideal image. It has been reported that at least 21 variations of this approach have been used as templates (Thompson and Gray, 1995). However, their reliability has largely been associated with the coarseness of the scale and limited choice, and the non-uniform interval between adjacent figures presents difficulties in analysis (Gardner *et al.*, 1998).

Further, a linear scale commonly addresses one aspect of body image, most commonly adiposity, and ignores muscularity or frame size, and, in reality, shape change is multidimensional. Gruber and co-workers avoided this difficulty by constructing a bi-axial matrix of photographic images of individuals representing a range of both muscularity and adiposity (Gruber *et al.*, 2000).

By contrast, distortion mirrors offer the advantage of the subject interacting with his/her own image. However, these have a non-linear response to adjustment which may provide implausible output which can be problematic to quantify. Electronic images of the individual were originally developed using analogue television technology, where adjustment would allow distortion in only X or Y planes. Guaraldi *et al.* (1995) compared a TV distortion method with a 'Cathexis scale' based on regional satisfaction ratings in a broad age range of female subjects. Further, TV distortion in horizontal and vertical scales fails to capture the likely phenotype relating to over or undernourishment, principally because it expands the skeletal, muscular and adipose tissue equally. The reality is that the skeletal dimensions do not alter appreciably with weight change (Davis *et al.*, 1994). A more sophisticated computer method was developed using algorithms, which created equal size changes between adjacent images on a 50-frame linear scale between two female subjects of BMI 18 and 42 (Stewart *et al.*, 2001). Despite its reliability, the instrument used a uni-axial approach to vary appearance, using images of two extremes and interpolation. This may fail to capture the complexity of phenotypic variation of the vast majority of individuals who fall in between.

More recently, digital photographic images have been used, which offer the advantage of being capable of manipulation via an interactive interface. A subject's own physique can been distorted via specialised computer algorithms which create a plausible and regionally specific effect. This technology was developed by Benson and co-workers and represented a breakthrough in the capacity for subject interaction, and quantification of both perception accuracy, and dissatisfaction (Benson *et al.*, 1999). Ideally, a method of assessment of body image should not influence body assessment by any individual. In reality, this is seldom the case, because the instrument fails to capture a credible image of the body, which is truly representative. However, by making the adjustment available in a user-friendly interface with digital images, the authors ensured such lack of transparency is minimised.

The purpose of this study was to investigate the relationship between physique and body image using this novel method in a sample, which embraces a wide range of size, adiposity and muscularity.

2 METHODS

2.1 Subjects

Sixty females and 77 males (mean age 27.7 +/− 9.7 y) representing a broad range of physique were used for the study. This sample was the combination of previously published data on recreational athletes (Stewart *et al.*, 2003), and unpublished data on healthy controls (Benson, unpublished data) and overweight/obese individuals (Johnstone *et al.*, unpublished data). Informed consent was obtained from each subject and ethical permission was granted by local regional and university research ethics committees.

2.2 Body Image Assessment

Each subject wore close-fitting exercise/swimwear and stood in a standard pose with arms abducted to 45°and legs approximately 0.5 m apart and was digitally photographed in the coronal plane. The camera-to-subject distance was 2.7–3.6 m, and adjustment for size was achieved using scaling factors provided by a 50 cm distance strip, which calibrated each image. Photographs were taken using a Kodak DCS260 (Kodak, Rochester, USA), a Toshiba PDR M61 (Toshiba Corporation, Taiwan) or a Sony Cybershot camera (Sony Corporation, Japan) in the three locations. Lighting was provided by twin 500W non-directional floodlights. Uncropped images were downloaded for image extraction from the background (Delin, VogueSoftware, UK, version 4.2b). This involved images being delineated using 106 manually-placed landmarks around the perimeter of the image. Delineated images were uploaded into the interactive programme (BodyImage, VogueSoftware, UK) which randomly distorted the image in nine body regions (left arm, right arm, chest, rib, hip, left thigh, right thigh, left calf and right calf) using algorithms based on perpendicular vectors originating from body segment mid-points. The image was presented on a computer screen with the head masked, and distortion was created via a tessellation map which produced a two-dimensional warping effect in the coronal plane. Subjects used interactive slider controls to re-create their perceived and desired images, which were quantified for pixel area difference from the actual image. Both for individual regions or the body as a whole, perception accuracy was defined as the difference between perceived and actual areas, and satisfaction was defined as the difference between the desired image and actual image areas. Pixel areas were converted into actual projected areas via use of appropriate factors derived from the 50 cm tape size in the image. These were the same for all subjects in each measuring location. Projected areas were calculated from the pixel area totals of all body regions except the head.

2.3 Anthropometry

Total body mass was measured to 100 g on a Seca Omega 873 digital floor scale (Hamburg, Germany) and stretch stature was obtained using a Holtain stadiometer (Crymych, UK). Skinfolds were measured by Harpenden Calipers (British Indicators, Luton, UK) at triceps, subscapular, supraspinale and medial calf sites. Girths were measured at the calf (maximum) and upper arm (flexed and tensed) using a modified Lufkin anthropometric tape (Rosscraft Innovations, Vancouver, Canada). Breadths were measured at the distal humerus and femur using a Tommy 2 bone caliper (Rosscraft Innovations, Vancouver, Canada). All anthropometric measurements made in triplicate by technicians were verified against those of a Level 3 or 4 ISAK anthropometrist following established guidelines (ISAK, 2001). Anthropometric somatotype was calculated by computer according to established formulae (Duquet and Carter, 2001). The total body surface area was calculated from stature and mass (DuBois and DuBois, 1916).

2.4 Statistical methods

Statistical analysis involved unpaired T tests to compare male-female differences in morphology and satisfaction. Correlation and regression analysis of dissatisfaction with separate somatotype components as predictor variables were performed using SPSS version 13 using the Bonferroni adjustment for multiple comparisons.

Results for regional dissatisfaction were calculated by summing left and right limb results. A whole-body cumulative dissatisfaction score was calculated by the square root of the sums of the squares of the regional dissatisfaction scores. Projected area was calculated via the sum of the undistorted regional areas, excluding the head.

3 RESULTS

Physical characteristics of the subjects are summarised in Table 16.1 and somatotypes of subjects are illustrated in Figure 16.1.

Table 16.1 Physical characteristics of subjects.

	Males (n = 77)		Females (n = 60)	
	Mean ± SD	Range	Mean ± SD	Range
Age (years)	27.3 ± 8.9	19–52	28.2 ± 10.6	18–55
Stature (cm)	178.3 ± 5.9	163.8–191.1	165.0 ± 7.5**	146.3–182.5
Body Mass (kg)	80.0 ± 14.2	52.8–125.1	64.6 ± 15.2**	41.1–114.9
BMI (kg.m-2)	25.1 ± 4.0	17.6–42.9	23.7 ± 5.2	17.3–39.2
Endomorphy	3.5 ± 1.6	1.4–8.6	4.8 ± 2.0**	1.1–9.4
Mesomorphy	5.5 ± 1.3	3.0–11.2	4.7 ± 2.1*	0.9–11.6
Ectomorphy	2.1 ± 1.1	0.1–5.3	2.2 ± 1.4	0.2–5.6

Significantly different: **P < 0.001 *P < 0.01 Somatotype attitudinal mean was 2.1 for males and 2.8 for females.

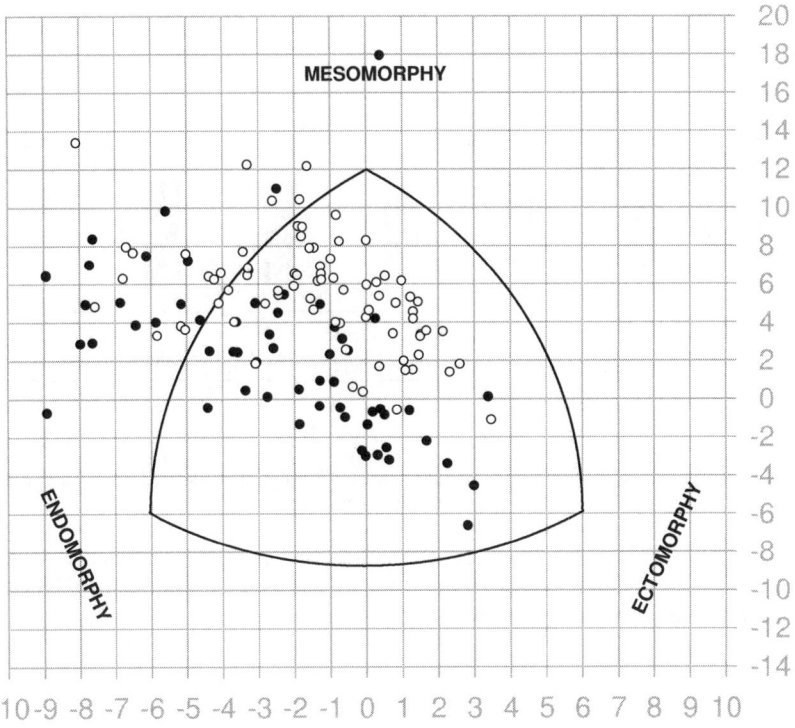

● Female   ○ Male

Figure 16.1 Somatotypes of subjects.

236   *Stewart* et al.

Reproducibility of the method for assessing perception and satisfaction previously established intraclass correlation coefficients of 0.88 and 0.89 respectively (P<0.01) (Stewart *et al.*, 2003). No differences were observed between perception accuracy for any region between male and females (P > 0.05), but there were considerable differences in satisfaction. Female subjects wished to be smaller in all body regions, averaging 5.6%, and ranging from 1.4% in the calf, to 9.4% in the thigh. Males wished to be larger in most regions, averaging 5.6% overall, and ranging from 1.3% smaller in the hip, to 10.7% larger in the calf. Regional dissatisfaction scores were significantly different between men and women for every body region (P < 0.001 after Bonferroni correction) and are illustrated in Figure 16.2. Analysis of the relationship between physique and dissatisfaction proceeded with separate analysis for males and females. Regional dissatisfaction scores were correlated with somatotype components and regressed against endomorphy, mesomorphy and ectomorphy as independent predictors.

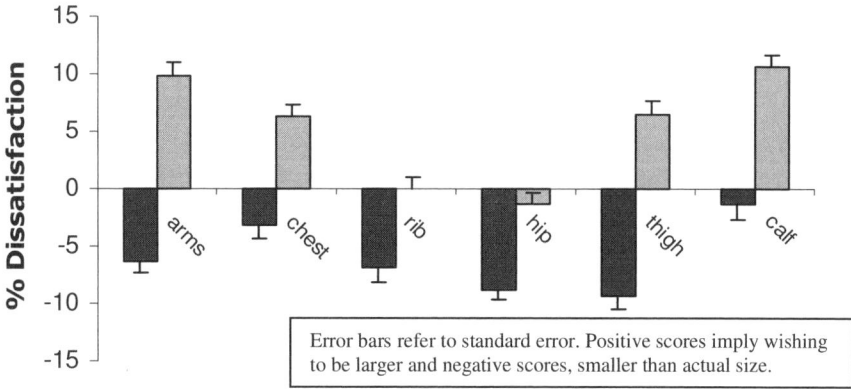

Figure 16.2 Regional dissatisfaction in males (light) and females (dark) defined as pixel area difference between ideal and actual.

The results of the analyses are summarised in Table 16.2 and Figure 16.3. Similarly, cumulative dissatisfaction scores were entered into stepwise regression analysis with the same predictor variables.

Table 16.2 Correlation between somatotype and regional satisfaction by sex.

	Endomorphy	Mesomorphy	Ectomorphy
Males n = 77			
Arm	−0.40**	−0.21	0.22
Chest	−0.57**	−0.28	0.32
Rib	−0.54**	−0.39**	0.37*
Hip	−0.44**	−0.44**	0.37*
Thigh	−0.54**	−0.45**	0.40**
Calf	−0.33*	−0.29	0.28
Cumulative dissatisfaction	0.14	0.13	−0.04
Females n = 60			
Arm	−0.65**	−0.53**	0.65**
Chest	−0.56**	−0.19	0.39*
Rib	−0.65**	−0.39*	0.56**
Hip	−0.49**	−0.22	0.41*
Thigh	−0.63**	−0.56**	0.71**
Calf	−0.51**	−0.46**	0.52**
Cumulative dissatisfaction	0.72**	0.50**	−0.55**

*P<0.05 (two tailed) ** P<0.01 (two tailed) after Bonferroni adjustment

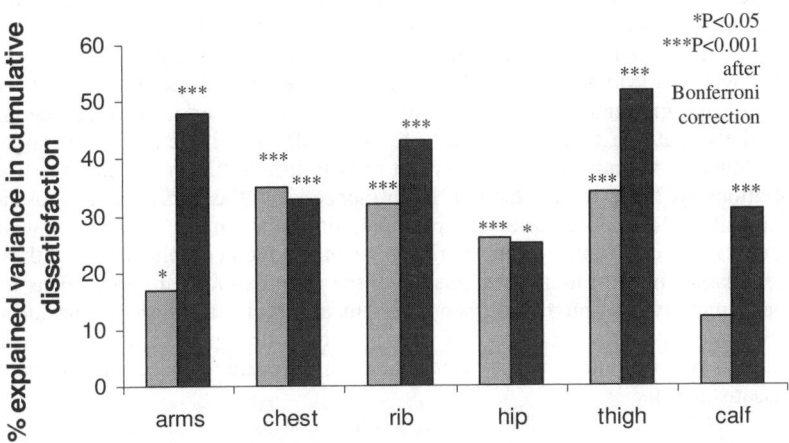

Figure 16.3 Explained variance in regional dissatisfaction by somatotype in males (light) and females (dark).

In males, no significant prediction for cumulative dissatisfaction was identified (P=0.26), while in females, somatotype explained 54% of the variance (P<0.001). Scrutiny of the regression results attributed almost all dissatisfaction to endomorphy, with mesomorphy a significant predictor in hip and thigh regions in males. Ectomorphy was not a significant predictor in any body region for either sex.

The projected area (excluding the head) and the total surface area (DuBois and DuBois, 1916) were divided by mass and used in regression analysis. In women, regression analysis of surface area divided by mass explained 54% of cumulative dissatisfaction. Projected area divided by mass explained 7% of the cumulative dissatisfaction ($P < 0.05$), but was not selected in a combined model using stepwise regression analysis. In males, neither surface area divided by mass, nor projected area divided by mass was a significant predictor of cumulative dissatisfaction ($P = 0.84$ and 0.14 respectively).

4 DISCUSSION

These data show a strong gender difference in regional dissatisfaction, which is explained partly by somatotype. In males, the correlation of regional dissatisfaction of rib, hip and thigh regions was significant for all somatotype components, and of arm, chest and calf regions for endomorphy only. The lack of a significant relationship between chest, arm and calf dissatisfaction with mesomorphy is surprising, and may relate to issues relating to muscle definition and shape, rather than the purely dimensional measures of the anthropometric somatotype. In females, dissatisfaction was significant for all areas with all somatotype components except the chest and hip for mesomorphy ($P = 0.14$ and $P = 0.09$ respectively). The poorer correlation with cumulative dissatisfaction with somatotype suggests either it is a poor summary of total dissatisfaction, or portrays a sense that when the dissatisfaction of some areas are effectively diluted by others, the overall meaning is lost. While a principal weakness of the body image assessment instrument is the failure to distinguish between fat and fat-free mass, it is likely that the significant levels of dissatisfaction in females can largely be attributed to variations in adiposity. At present it is not known if the mesomorphy relationship with dissatisfaction is influenced by the presence or absence of adiposity, and a more varied sample, including more muscular subjects would be necessary to clarify this. Males perhaps have the added complication that they wish to be larger in most areas, but smaller in the abdomen and hip areas. This renders the cumulative dissatisfaction score less meaningful than those of the individual body regions. However, one of the principal strengths of the study include minimising the potential for size estimation inaccuracy to be misinterpreted as dissatisfaction.

While males and females view their bodies in radically different ways, dissatisfaction studies have rarely considered an objective physique assessment. The short paper by Calden (1959) is an exception, where seven somatotype photographs from the Atlas of Men by Sheldon *et al.* (1954) showed the central somatotype to be the most desirable, with the extreme endomorph being the least desirable. The range in physique represented by this sample has enabled a far more comprehensive and subject-specific analysis of dissatisfaction in both sexes. Despite a potentially infinite number of test outcomes, the test–retest results are in sufficiently good agreement to provide a valid measurement tool. One of the criticisms of the reported high construct validity of other instruments is that high test–retest scores are probably a function of limited response options for the observer (Gardner *et al.*, 1998).

Postural considerations involving the precise degree of shoulder or hip rotation could have affected the projected area to an extent which has not been quantified. Clothing worn for the photography was standardised as far as reasonably possible using the close-fitting Lycra sportswear provided for the experiments. However, overweight or obese subjects were required to wear swimwear as part of another study, and wore this for the photography. In some cases delineation proved problematic where the outline of the wearer's shape (e.g. abdominal panniculus) was not readily apparent. It is of some surprise that projected area did not correlate more strongly with dissatisfaction, perhaps because subjects did not include many extremely-obese subjects whose projected area/mass relationships would differ. It is probable that more extreme forms of obesity might affect a sagittal more than a coronal plane projected area. Despite being nearly a century old, the predicted surface area calculation from DuBois and DuBois (1916) is still in current use. The limited sample size, and the application of a formula to both men and women perhaps calls into question its use today. However, it did explain a significant degree of variation in dissatisfaction in women, perhaps because a higher mass would involve a higher fat content, which was not necessarily true in males.

The variability of the human physique is considerable and it appears clear that males and females express different levels of dissatisfaction with different body regions. This may not be surprising, but this is the first study to quantify its extent in such a variable sample. The reality for some is that the desirable physique may be physiologically implausible, which, once this is realised over time, may trigger either an accumulation or accommodation of the burden of the dissatisfaction (McLaren and Gauvin, 2002). In addition, a desirable physique may be psychologically unachievable, as the value judgements of what constitutes ideal may change with time, as does the capacity to alter the body shape in response to diet and exercise. A textural component may become relevant in older individuals where dissatisfaction relates to a sensation of firmness (or flabbiness), which accompanies the dimensional measure.

The utility of this field of research could be enhanced if photoscopic somatotype were calculated in addition to anthropometric somatotype. The selectivity involved in summarising any physique from 10 anthropometric measures has consequences for the interpretation of regionally specific data on dissatisfaction. Regional dysplasia resulting from fat accumulation or specific muscle development necessitates adjustment of the 'true somatotype' and this is only possible using experienced raters of the photoscopic method. The sagittal plane perspective – a part of the photoscopic somatotype procedure, would be a useful adjunct to the method performed in this study, and would enable greater linkage with physique variation in any individual subject. Further work quantifying projected and total surface area using digital techniques needs to be developed and refined to minimise errors. In this respect, 3D laser scanning offers considerable potential not only for dimensional measures, but for quantification of curved surface geometry and providing an enhanced visualisation tool to assist participants view a more complete representation of their body image.

5 CONCLUSION

In conclusion, males and females display widely different body dissatisfaction ratings which are explained only partly by somatotype, more strongly in females than males. The inability of projected area to explain dissatisfaction to any great extent is likely to reflect difficulties in landmarking accuracy or greater phenotypic variation than depicted in a coronal plane.

6 REFERENCES

Benson, P.J., Emery, J.L., Cohen-Tovée, J. and Tovée, M.J., 1999, A computer-graphic technique for the study of body size perception and body types, *Behaviour Research Methods, Instruments and Computers*, **31**, 446–454.

Calden, G. (1959). Sex differences in body concepts. *Journal of Consulting Psychology*, **23**, 378.

Davis, C., Durnin, J.V.G., Dionne, M. and Gurevich, M., 1994, The influence of body fat content and bone diameter measurements on body dissatisfaction in adult women. *International Journal of Eating Disorders*,**14**, 257–263.

Dubois, D. and DuBois, E., 1916, Clinical Calorimetry. A formula to estimate the approximate surface area if height and weight be known. *Archives of Internal Medicine*, **17**, 863–871.

Duquet, W. and Carter, J.E.L., 2001, Somatotyping. In *Kinanthropometry and Exercise Physiology Laboratory Manual: Tests, Procedures and Data*, 2nd edition Volume 1: Anthropometry (Edited by R.G. Eston and T. Reilly), pp. 47–64. London: Routledge.

Gardner, R.M., Friedman, B.N. and Jackson, N.A., 1998, Methodological concerns when using silhouettes to measure body image. *Perceptual and Motor Skills*, **86**, 387–395.

Gruber, A.J., Pope, H.G. Jr., Borowiecki, J.J. and Cohane, G., 2000, The development of the somatomorphic matrix: A biaxial instrument for measuring body image in men and women. In *Kinanthropometry VI* (Edited by K. Norton, T. Olds and J. Dollman), pp. 217–232. Underdale, Australia: ISAK.

Guaraldi, G.P., Orlandi, E., Boselli, P. and Tartoni, P.L., 1995, Body size perception and dissatisfaction in female subjects of different ages. *Psychotherapy and Psychosomatics*, **64**, 149–155.

International Society for the Advancement of Kinanthropometry (ISAK), 2001, *International Standards for Anthropometric Assessment*, Underdale, SA, Australia.

Kay, S., 1996, The psychology and anthropometry of body image. In Anthropometrica (Edited by K. Norton and T. Olds), pp. 236–258. Sydney: University of New South Wales Press.

McLaren, L. and Gauvin, L., 2002, The cumulative impact of being overweight on women's body esteem: A preliminary study, *Eating and Weight Disorders*, **7**, 324–327.

Sheldon, W.H., Dupertuis, C.W. and McDermott, E., 1954, *Atlas of Men*, New York: Harper and Brothers.

Stewart A.D., Benson, P.J., Michanikou, E.G., Tsiota D.G. and Narli, M.K., 2003, Body image perception, satisfaction and somatotype in male and female athletes and non-athletes: Results using a novel morphing technique. *Journal of Sports Sciences*, **21**, 815–823.

Stewart, T.M., Williamson, D.A., Smeets, M.A.M. and Greenway, F.L., 2001, Body morph assessment: Preliminary report on the development of a computerised measure of body image. *Obesity Research*, **9**, 43–50.

Thompson, J.K., 1990, *Body Image Disturbance: Assessment and Treatment*, New York: Pergamon.

Thompson, M.A. and Gray, J.G. (1995). Development and validation of a new body image assessment scale. *Journal of Personality Assessment*, **64**, 258–269.

CHAPTER SEVENTEEN

Proportionality and sexual dimorphism in elite South African crawl stroke swimmers

B. Coetzee

North-West University, Potchefstroom, 2520, South Africa

1 INTRODUCTION

The direct relationship between a swimmer's anthropometric composition and his/her potential to succeed has long been recognized by researchers in sport science (Montpetit and Smith, 1988; Ackland and Mazza, 1994; Pelayo *et al.*, 1996). Research also seems to suggest that swimmers develop distinct morphological attributes with training and participation in swimming (Siders *et al.*, 1991; Ackland and Mazza, 1994) which may optimize their swimming potential. The unisex Phantom is used in this regard to assess the anthropometric proportionality characteristics of a certain population of athletes compared to that of a normal population (Ross and Marfell-Jones, 1991). The unique anthropometric characteristics that are not commonly observed in the general population will, therefore, be identified and may be used in talent identification programmes.

A few studies (Ross *et al.*, 1981, 1982; Ross and Ward, 1984) have made use of the proportionality approach to try and determine the anthropometric characteristics which distinguish swimmers from the general population. These studies have collectively shown that Olympic swimmers of 1960, 1968 and 1976 possessed proportionally smaller fat mass and skinfold values, higher muscle mass values, higher hand and upper arm lengths, larger anterior-posterior chest depths, smaller biiliocristal breaths and biepicondylar femur widths, larger relaxed arm and chest girths, smaller thigh and calf girths as well as lower body weight values than the Phantom. A comparison between the absolute body size characteristics of swimmers of the Montreal Olympic Games (1976) (Carter *et al.*, 1982) and the 1991 World Championships of Swimming (Mazza *et al.*, 1994) clearly showed that swimmers had evolved significantly in terms of their anthropometric composition during the 15-year time span. In view of this, the relevancy of swimmers' proportional data that are 20 to 30 years old to today's swimmers can be questioned. Furthermore, it has also been unclear whether elite South African swimmers would also exhibit the same type of trends with regard to their anthropometric proportionality as do their international counterparts.

The Phantom has also been used to superimpose male and female profiles in the same sport to examine sexual dimorphism among a population of athletes who participate in the same type of item and training regimens (Ross *et al.*, 1982; Carter and Ackland, 1998; De Ridder *et al.*, 2003). The term 'sexual dimorphism' refers to any differences in form between the sexes in a certain population of individuals (Pinkstaff, 1998). As expected, the existing literature shows that striking differences exist between men and women with regard to a wide range of variables, namely: anthropometric measurements (Withers *et al.*, 1998; Nindl *et al.*, 2002), performances parameters such as strength (Bishop *et al*, 1987; Shephard, 2000), flexibility (Bell and Hoshizaki, 1981), ventilatory responses to exercise (Kilbride *et al.*, 2003), as well as physiological variables such as circulating endogenous hormones (Gatford *et al.*, 1998; Rosenbaum and Leibel, 1999). The occurrence of sexual dimorphism among elite swimmers is, however, a research area that has not received as much attention during the last few years. The research that does exist will, however, be discussed in the next section to gain a clearer understanding of the extent of anthropometric dimorphism among elite swimmers.

Vervaecke and Persyn (1981) used 47 successful Belgium national swimmers of both genders in their study and found that female swimmers showed significantly lower values for hand ($\overline{X}$ = 0.21 m^2 compared to 0.23 m^2) and foot surface areas ($\overline{X}$ = 0.239 m^2 compared to 0.267 m^2) as well as arm ($\overline{X}$ = 42.85 cm compared to 44.78 cm) and leg lengths ($\overline{X}$ = 46.58 cm compared to 48.54 cm) when they were compared to the male swimmers. The majority of research results concerning gender differences in somatotype among swimmers demonstrated that female swimmers fall into the central category of 3-4-3 compared to the male swimmers who usually fall into the ectomesomorph category of 2-5-3 (Coetzee, 2002, Carter and Marfell-Jones, 1994). These findings show that the endomorphy values of female swimmers are generally higher and their mesomorphy values generally lower than those of male swimmers. Similar results with regard to the differences between female and male swimmers' somatotypes were also reported by Gualdi-Russo and Graziani (1993) in their study of Italian swimmers. Studies on Olympic and South African swimmers respectively did, however, show that female swimmers obtained higher ectomorphy values compared to the male swimmers (Carter, 1984; Coetzee, 2002).

The literature further indicated that successful female swimmers showed lower average arm span, sitting height, girth, breadth and length values than their male counterparts (Ross et al., 1994). Moreover, female World Championship swimmers demonstrated smaller values in all body composition variables (body mass, percentage muscle mass and percentage skeletal mass) compared to the male swimmers, except with regard to the sum of six skinfolds where the female group obtained the highest average value (Drinkwater and Mazza, 1994). McLean and Hinrichs (2000) also noted that that the nineteen female swimmers in their study had significantly more body fat (24.1%) than the thirteen male swimmers (14.8%) that were measured. These findings are also supported by Simmons (2003) who found that female and male college-age swimmers were significantly different on all measured variables, with males being taller, heavier and leaner than the females.

The world records by female swimmers are, on average, 6–10% behind those of male swimmers (Åstrand *et al.*, 2003). Previous research has shown that male arm speed (stroke frequency) is one of the major contributors to their being faster than female swimmers (Smith, 1978). A more recent study by Takagi *et al.* (2004) found that stroke length was significantly greater in male compared to female swimmers. Grimston and Hay (1986) indicated that arm and leg length, hand cross-sectional area, leg frontal area and foot cross-sectional area significantly correlated with stroke frequency and length among swimmers. These results suggest that differences in the stroke indices and swimming speeds of female and male swimmers were the direct consequence of the differences in their anthropometric make up. This notion is further supported by Simmons (2003) who reported that the greater muscularity of male swimmers can be utilized to enhance strength, power and ultimately swim performance.

There would also appear to be gender-specific responses of plasma cortisol and testosterone concentrations to different endurance training regimens among swimmers (Tyndall *et al.*, 1996). Support for this view was provided by Wilmore and Costill (2004) who observed that females generally gain substantially less muscle mass in response to a given training stimulus than males do, probably because of their lower levels of testosterone. The effect of sexual dimorphism on the response to different swim training workloads and regiments can, therefore, not be questioned. Gualdi-Russo and Graziani (1993) did, however, speculate that greater sexual dimorphism will be observed in athletes who participate in different activities during training and competitions than athletes where both genders perform the same activities during the last mentioned time periods.

It is against this background and the lack of research concerning the link between swimming and the anthropometric proportionality as well as the incidence of sexual dimorphism with regard to the anthropometric variables, that this study was undertaken. The purpose of this study was, therefore, twofold: firstly, to establish the absolute size and anthropometric proportionality characteristics of elite South African crawl stroke swimmers. Secondly, to determine the extent of sexual dimorphism in elite South African crawl stroke swimmers. Information that arises from this study may provide important knowledge to sport professionals with regard to the relationship between swimming and sexual dimorphism as well as establish a link between participation in swimming and the effect of this on the development of certain anthropometric traits among swimmers.

2 METHODS

2.1 Research design

The design of the study was a cross-sectional, experimental design. Information was obtained by means of a questionnaire and anthropometric measurements. The questionnaire consisted out of four main sections, namely: General information, information regarding training habits, medical information and competition data. Each of the sections contained questions that were related to the particular section heading. The objective of the study was explained to the swimmers, after which they all completed informed consent forms.

2.2 Subjects

Sixteen male (18.69 ± 3.28 years) and ten female (17.0 ± 2.36 years) swimmers who finished in the ten top positions of the different crawl stroke events at the South African Swimming Championships were measured in this study.

2.3 Anthropometric measurements

A total of 41 anthropometric variables were directly and indirectly measured according to the methods of Norton *et al.* (1996a) and included body mass, skinfolds, girths, breadths, lengths, heights and selected body composition components. Specific variables are identified in the Tables and Figures. Another ten body composition measures were indirectly determined and included: lean body mass, body density, sum of the six skinfolds, fat mass and percentage, muscle mass and percentage as well as the somatotype components which consisted of the endomorphy, mesomorphy and ectomorphy values. Somatotype was calculated using the formula of Carter and Heath (1990), fat percentage by using the formulas of Withers *et al.* (1987a,b) and muscle mass by using the formula of Martin *et al.* (1990).

2.4 Statistical analysis

The Statistica Data Processing package (StatSoft Inc., 2005) was used to process the data. Firstly, the descriptive statistics of each anthropometric variable were calculated. Secondly, independent t-tests were used to determine the statistically significant differences ($p < 0.05$) between the anthropometric measurements of the male and female swimmers. This was followed by the calculation of the unisex Phantom Z-values which were used to determine the proportional profile of the swimmers and to compare the two genders of swimmers in a proportional manner. [The Phantom is a calculation device and a hypothetical, unisex reference human with defined p values for over 100 anthropometric measurements which has the ability to quantify proportional differences in anthropometric characteristics between subjects (Ross and Marfell-Jones, 1991).] The Phantom values were graphically displayed as proportionality profiles. Z-values of 0 indicate that a score is proportionally the same as the Phantom. Positive Z-values indicate that the particular measure is proportionally larger than that of the Phantom, while negative values show that the measure is proportionally smaller (Ross and Ward, 1984). Lastly, statistical differences between the two groups of swimmers were calculated by comparing the standard errors about the mean of each population. Standard error bars about a sample mean that did not overlap those of another mean, were deemed significant at the 5% probability level (Ross and Marfell-Jones, 1991).

3 RESULTS

Firstly, the descriptive statistics and the significance of the different independent t-test results for each of the absolute anthropometric variables with regard to the different gender groups are presented in Tables 1 to 6.

Table 17.1 shows the descriptive statistics of the skinfold variables as well as the independent t-test results of the differences between the gender groups with regard to the these variables. The triceps, subscapular, supraspinal, front thigh and medial calf skinfolds were the variables in Table 17.1 which showed significant differences between the two gender groups. In spite of the fact that the average abdominal skinfold obtained a much higher value for the females (15.7 mm) compared to the males (11.9 mm) no significant difference was observed. The females obtained higher average skinfold values in all of the measurements.

Table 17.1 Descriptive statistics and significance of the different independent t-test results for the skinfold variables with regard to the different gender groups.

Variables	Females (n = 10)				Males (n = 16)			
	Mean	Min	Max	SD	Mean	Min	Max	SD
Triceps (mm)*	12.7	10.1	18.2	2.6	8.5	5.4	11.5	2.0
Subscapular (mm)*	9.4	7.5	13.1	1.7	8.0	5.8	10.8	1.6
Pectoral (mm)	5.9	4.1	7.7	1.4	5.7	3.5	10.3	1.8
Abdominal (mm)	15.7	9.2	27.0	6.0	11.9	6.9	20.0	4.0
Iliac crest (mm)*	15.2	6.9	24.1	5.6	11.4	7.4	16.6	3.2
Supraspinale (mm)*	10.5	6.4	15.8	3.2	7.5	4.2	13.6	2.4
Front thigh (mm)*	24.3	12.7	34.9	7.9	13.2	7.7	18.4	3.2
Medial calf (mm)*	14.3	9.0	20.0	4.2	9.8	5.0	15.8	2.8

SD = Standard deviation

Min = Minimum values

Max = Maximum values

* Females and males significantly different (p < 0.05)

Table 17.2 contains the descriptive statistics of the girth variables as well as the independent t-test results of the gender differences with regard to these results.

The results in Table 17.2 show that the males obtained larger average values in all the girth measurements except two, namely gluteal and thigh girths. These were, however, also the only girth variables for which no significant differences between the two genders were shown.

The descriptive statistics of the length variables and the independent t-test results of the gender differences with regard to these variables are shown in Table 17.3.

Table 17.2 Descriptive statistics and significance of the different independent t-test results for the girth variables with regard to the different gender groups.

Variables (cm)	Females (n = 10)				Males (n = 16)			
	Mean	Min	Max	SD	Mean	Min	Max	SD
Head*	54.9	52.8	56.0	1.1	56.6	54.3	58.5	1.2
Neck*	32.4	30.0	34.0	1.3	37.5	34.0	41.5	2.3
Relaxed arm*	27.4	24.7	30.3	1.9	31.0	26.5	35.5	2.5
Flexed arm*	28.6	25.7	32.3	2.0	33.2	29.4	37.0	2.4
Forearm (relaxed)*	24.0	22.7	25.9	1.2	27.0	20.2	31.0	2.5
Wrist (distal styloids)*	15.3	14.6	16.2	0.5	17.3	16.3	19.3	0.8
Chest (mesosternal)*	90.0	83.0	98.5	4.8	100.5	91.0	111.2	6.4
Waist (minimum)*	69.9	61.4	76.4	4.2	78.2	70.9	86.5	4.9
Gluteal (hips)	94.4	86.8	100.8	3.8	93.3	86.1	103.4	5.9
Thigh (1cm distal gluteal line)	57.0	50.7	62.0	3.0	55.9	50.7	63.1	3.8
Calf (standing)*	34.3	31.7	38.3	1.9	37.0	33.9	44.9	3.1
Ankle*	21.4	20.2	22.7	0.8	22.9	20.5	24.9	1.4

SD = Standard deviation

Min = Minimum values

Max = Maximum values

* Females and males significantly different (p < 0.05)

Table 17.3 Descriptive statistics and significance of the different independent t-test results for the length variables with regard to the different gender groups.

Variables (cm)	Females (n = 10)				Males (n = 16)			
	Mean	Min	Max	SD	Mean	Min	Max	SD
Body stature*	170.5	163.0	177.4	4.5	182.6	173.7	192.6	5.8
Sitting height*	87.9	84.0	92.9	3.0	92.7	86.3	99.4	3.5
Arm span*	174.0	169.3	183.3	4.2	189.2	170.3	203.2	9.3
Arm (acromiale-radiale)*	32.6	30.9	34.5	1.3	35.3	30.5	38.6	2.0
Forearm (radiale-stylion)*	25.5	24.2	27.1	1.0	27.8	25.8	30.6	1.6
Hand (stylion-dactylion)*	18.4	18.0	19.0	0.3	19.9	18.5	22.5	1.2
Thigh (trochanterion-tibiale)*	44.3	42.0	48.7	2.1	46.8	42.4	52.5	2.5
Tibial (lateral to floor)*	44.8	42.5	47.1	1.5	49.9	45.2	56.8	2.6
Tibia (tibiale med-sphyrion tib)*	37.7	35.1	40.0	1.3	40.5	37.1	43.6	1.9
Foot (akropodion-pternion)*	24.6	22.9	25.8	1.0	27.1	25.1	29.5	1.2

SD = Standard deviation

Min = Minimum values

Max = Maximum values

* Females and males significantly different (p < 0.05)

From Table 17.3 it is clear that the males obtained significantly larger average values in all the length variables compared to the females.

The descriptive statistics of the breadth variables together with the independent t-test results of the gender differences with regard to these results are presented in Table 17.4.

Table 17.4 Descriptive statistics and significance of the different independent t-test results for the breadth variables with regard to the different gender groups.

Variables (cm)	Females (n = 10)				Males (n = 16)			
	Mean	Min	Max	SD	Mean	Min	Max	SD
Biacromial *	38.8	37.2	40.9	1.2	41.9	32.0	48.3	3.8
Biiliocristal	28.4	24.9	37.9	3.7	28.9	25.3	31.6	1.9
Transverse chest (mesosternal) *	27.5	25.5	29.7	1.3	31.0	19.3	36.0	3.7
Chest depth (AP, mesosternal) *	17.5	13.8	20.0	2.0	20.7	14.7	29.7	3.1
Biepicondylar (humerus)*	6.3	5.7	6.7	0.3	7.2	6.5	7.8	0.4
Hand (distal II-V metacarpals)*	7.6	7.0	8.4	0.4	8.6	7.5	9.4	0.5
Biepicondylar (femur)*	8.9	8.4	9.3	0.3	9.6	8.7	10.5	0.5
Foot (distal I-V metatarsal)*	8.9	8.5	9.8	0.5	9.8	9.2	10.8	0.6
Bideltoid*	42.9	40.1	45.0	1.7	47.8	40.5	54.5	3.2
Bitrochanteric	32.0	29.4	33.9	1.3	32.4	29.6	37.1	2.1

SD = Standard deviation

Min = Minimum values

Max = Maximum values

* Females and males significantly different (p < 0.05)

Again Table 17.4 shows that the males obtained significantly larger average breath measurements in all the variables except two, namely biiliocristal and bitrochanteric breaths in which they obtained non-significant larger values.

The male swimmers' measurements were significantly larger than the females' measurements in six of the variables, namely: body mass, lean body mass, density, muscle mass, muscle mass percentage and mesomorphy (see Table 17.5). In contrast with these results the females obtained significantly larger average values in the following body composition variables: sum of six skinfolds, fat mass and endomorphy. Ectomorphy was the only body composition variable that did not show a significant difference between the two gender groups.

Table 17.5 Descriptive statistics and significance of the different independent t-test results for the body composition variables with regard to the different gender groups.

Variables	Females (n=10)				Males (n=16)			
	Mean	Min	Max	SD	Mean	Min	Max	SD
Body mass (kg)*	61.6	53.3	72.8	5.4	74.7	60.4	92.8	10.1
Lean body mass (kg)*	49.3	44.2	57.6	3.7	67.1	55.5	81.3	7.9
Density (g/cm3)*	1.05386	1.04139	1.06293	0.00749	1.07628	1.06400	1.08521	0.00518
Sum of the six skinfolds (mm)*	86.7	58.1	123.3	22.3	58.9	35.5	88.4	13.3
Fat mass (kg)*	12.2	8.6	16.5	2.8	7.6	4.5	13.3	2.6
Fat percentage (%)*	19.7	15.7	25.3	3.4	9.9	6.1	15.2	2.2
Muscle Mass (kg)*	29.6	26.9	35.2	2.9	40.9	30.4	50.6	6.6
Muscle percentage (%)*	48.2	43.9	52.1	3.5	54.7	45.9	59.8	3.6
Endomorphy*	3.3	2.4	4.4	0.7	2.2	1.3	3.3	0.6
Mesomorphy*	3.3	2.0	4.5	0.9	4.4	3.1	5.9	0.9
Ectomorphy	3.1	1.6	4.8	1.1	3.3	1.5	4.7	1.0

SD = Standard deviation

Min = Minimum

Max = Maximum

*Females and males significantly different ($p < 0.05$)

From the data in Table 17.1–17.5 it is clear that significant differences were found in almost all of the absolute anthropometric variables between the two genders. Only seven of the possible 51 variables did not show significant differences between the two groups of swimmers. These included: pectoral and abdominal skinfolds; gluteal and thigh girths; biiliocristal and bitrochanteric breadths as well as ectomorphy.

Subsequently, the unisex Phantom Z-values were calculated and used firstly to determine the anthropometric proportionality profiles of the swimmers and secondly to compare the two genders of swimmers in a proportional manner. The Z-values that were obtained are presented graphically as proportionality profiles in Figures 17.1 to 17.5.

Figure 17.1 contains the Z-values and standard errors for the skinfold variables of the male and female crawl stroke swimmers. All the skinfold variables showed proportionally smaller values for the swimmers when compared to the

Phantom. Furthermore, the female swimmers obtained proportionally larger skinfold measurements compared to the males. The triceps, supraspinale, front thigh and medial calf skinfolds were, however, the only skinfold variables that displayed significant proportional differences between the two gender groups.

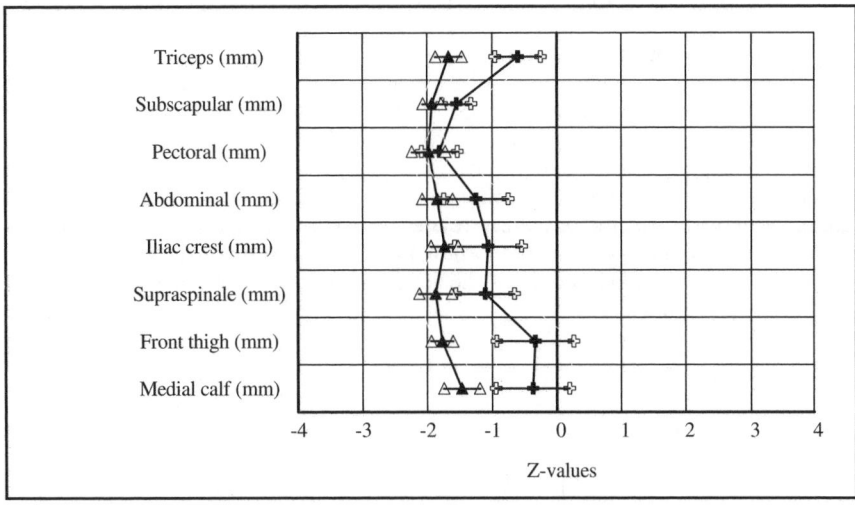

Figure 17.1 Proportionality profile of the Z-values (closed markers) and standard errors (open markers) for the skinfold variables of the male (open and closed triangles) and female crawl stroke swimmers (open and closed plusses).

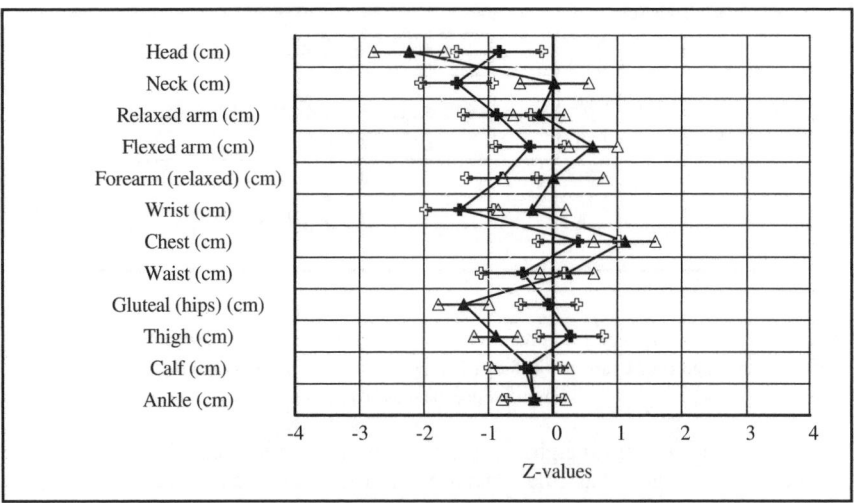

Figure 17.2 Proportionality profile of the Z-values (closed markers) and standard errors (open markers) for the girth variables of the male (open and closed triangles) and female crawl stroke swimmers (open and closed plusses).

Figure 17.2 displays the proportional girth results of the female and males swimmers, respectively.

From the results in Figure 17.2, it is clear that female swimmers generally displayed proportionally smaller girth measurements compared to the Phantom. Only chest and thigh girth obtained proportionally larger average values compared to the rest of the girth measurements which all displayed proportional smaller values. Similarly, the male swimmers possessed proportionally smaller girth values in seven out of a possible twelve girth variables. Male swimmers did, however, possess proportionally similar average neck and relaxed forearm girth values to the Phantom. The males were proportionally larger in flexed arm, chest and waist girths. The separation in standard error bars shows that the females compared to the males were proportionally significantly larger in head, wrist, gluteal and thigh girths. Neck girth was the only girth related variable in which the male swimmers displayed a significantly larger proportional average value than the females.

The proportionality scores of the length and height variables are displayed graphically in Figure 17.3.

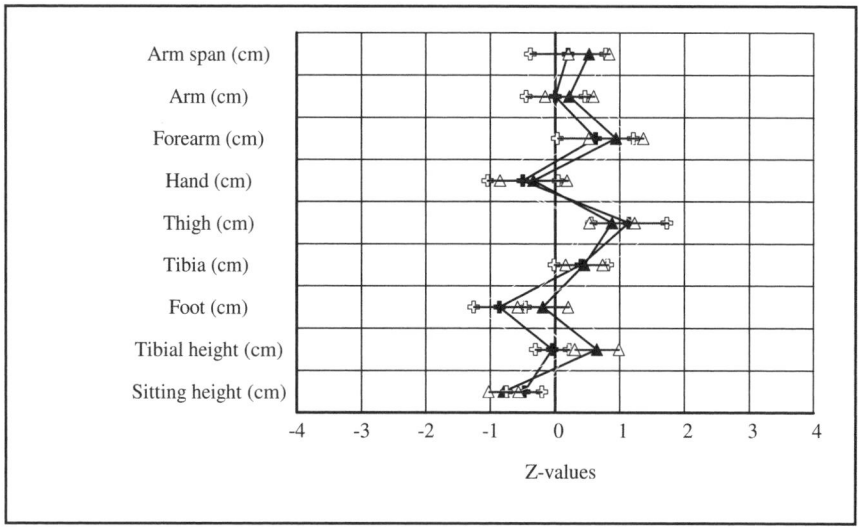

Figure 17.3 Proportionality profile of the Z-values (closed markers) and standard errors (open markers) for the length and height variables of the male (open and closed triangles) and female crawl stroke swimmers (open and closed plusses).

The length and height variables showed that the swimmers were proportionally larger than the Phantom in four variables namely: arm span, forearm, thigh and tibia lengths. They also displayed proportionally smaller hand and foot lengths as well as sitting height than the Phantom. Female swimmers had proportional smaller average arm lengths and tibial heights in contrast to male swimmers who had proportional larger values in the last-mentioned variables compared to the Phantom. However, female swimmers displayed a proportionally larger average thigh length and sitting height than the male swimmers. Male swimmers possessed proportional larger average values in the rest of the measurements compared to the females. The differences between the two genders were, however, not large enough to be judged significant.

The Z-scores for the ten breadths are illustrated in Figure 17.4.

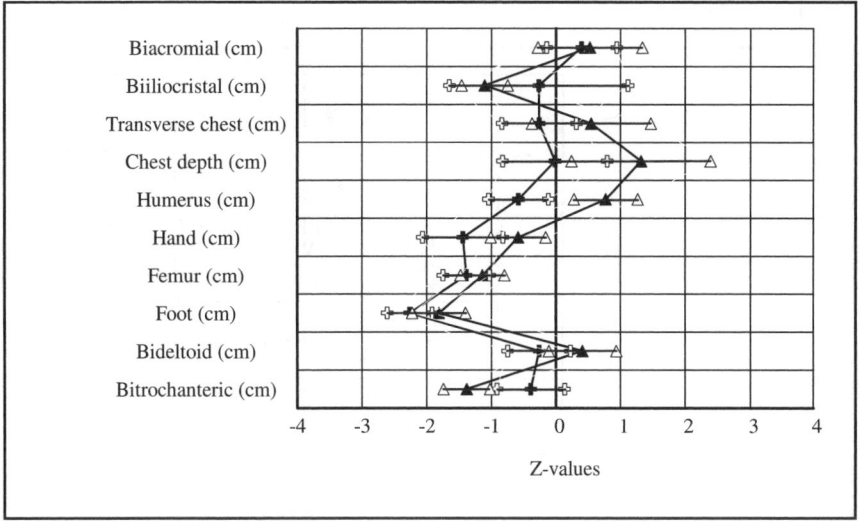

Figure 17.4 Proportionality profile of the Z-values (closed markers) and standard errors (open markers) for the breadth variables of the male (open and closed triangles) and female crawl stroke swimmers (open and closed plusses).

As shown in Figure 17.4, the female swimmers displayed smaller proportional breadth values in all the variables except one (biacromial breadth) in a comparison between their average values and the Phantom scores. The male swimmers, on the other hand, displayed smaller proportional breadth average values in five variables (biiliocristal, hand, femur, foot and bitrochanteric breadths) and larger values in five values (biacromial, transverse chest, chest depth, humerus and bideltoid breadths) compared to the Phantom. The proportional comparison between the two gender groups revealed that male swimmers possessed the larger breadths in most of the variables (eight out of a possible ten) when compared to the female swimmers. The biacromial and bitrochanteric breadths were the only variables in which the female swimmers obtained proportional larger average values than the male swimmers. Only two of the breadth variables (humerus and bitrochanteric breadths) did, however, show significant differences between the two genders.

The proportional profile for the body composition variables of the swimmers are presented in Figure 17.5.

Compared to the Phantom, the female swimmers were proportionally lighter in body mass and lean body mass, proportionally heavier in muscle mass, of proportional similar weight in fat mass and displayed a higher average proportional fat percentage (Figure 17.5). The male swimmers were proportionally lighter in body and fat mass, heavier in lean body and muscle mass and displayed a much lower average proportional fat percentage than the Phantom (Figure 17.5). In comparing the female and male swimmers, as shown in Figure 17.5, the female swimmers were proportionally heavier in body and fat mass, lighter in lean body and muscle mass and displayed a much higher average proportional fat percentage than the males.

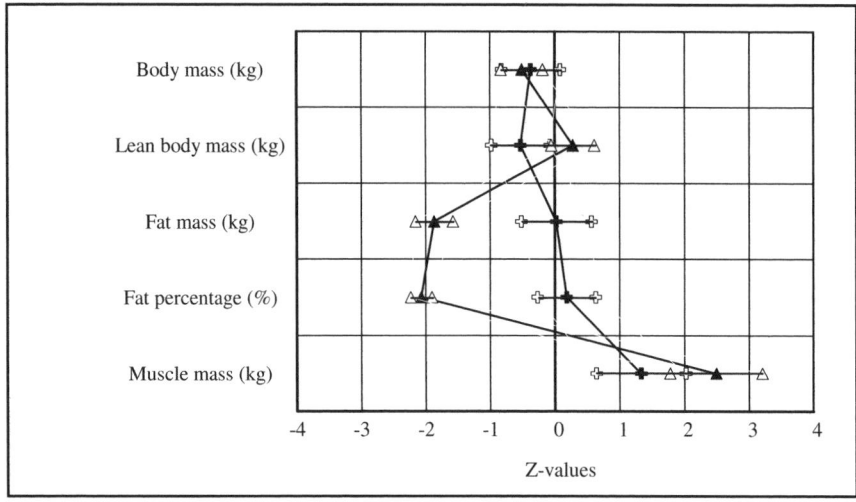

Figure 17.5 Proportionality profile of the Z-values (closed markers) and standard errors (open markers) for the body composition variables of the male (open and closed triangles) and female crawl stroke swimmers (open and closed plusses).

The only variables in which significant differences were found, were fat mass and percentage body fat.

The results of Figure 17.1 to 17.5 clearly show that the profiles for swimmers of both genders are very similar in appearance. This indicates that training and/or self-selection have a definite influence on elite South African swimmers' anthropometric proportionality. Furthermore, compared to the Phantom the swimmers were collectively proportionally larger in chest girth, longer in armspan, arm, forearm, thigh and tibia length, larger in biacromial breadth as well as in the muscle mass value. Compared to the Phantom, the female swimmers were also larger in thigh girth, fat mass and fat mass percentage. Proportional differences between male swimmers and the Phantom showed that the swimmers were proportionally larger in waist girth, transverse chest, chest depth, humerus and bideltiod breadth, longer in tibial height and larger in lean body mass. The males did, however, display similar proportional neck and forearm girths than the Phantom. Smaller (negative) values were displayed in all the other skinfold, girth, length, height, breadth and body composition variables when both swimmer groups were proportionally compared to the Phantom.

4 DISCUSSION

The finding that swimmers collectively display larger proportional chest girths corresponds with the results of Lafortuna and Passerini's study (1995) who found a direct, positive relationship between spirometer measurements and chest wall circumferences. A bigger chest girth may, therefore, be related to a better lung function. In this regard Kokkalis *et al.* (2001) demonstrated that lung function directly correlates with swimming performance among crawl stroke swimmers which can serve as an explanation for the larger chest girth. Additionally, swimmers who generally have larger upper limbs (longer armspan, arm and forearm lengths) are better suited for high power output due to the fact that the larger pulling cross-sectional area can be used to increase swimming propulsion (Jiang, 1993; Norton *et al.*, 1996b). The result that thigh and tibia length were also identified as anthropometric variables which yielded proportional higher values in the swimmers is not surprising in view of the fact that an increase in the cross-section area of limbs will result in an increase in water resistance against the specific limb so that effective propulsion can take place. This does not only apply to the arm stroke during swimming but also to the propulsion force of the legs. Research by Watkins and Gordon (1983) as well as Hollander *et al.* (1988) contradicts the prevailing opinion of some of the swimming coaches that the legs do not contribute to propulsion during the crawl stroke swimming events. They found that the kick contributed an average of 10–12% to propulsion during full-stroke swimming, which accentuates the importance of long lower limb lengths among crawl stroke swimmers.

According to Jiang (1993) wide shoulders (biacromial breadths) are the characteristic of most elite swimmers. Geladas *et al.* (2005) also reported a significant correlation ($r = 0.61$, $p < 0.01$) between shoulder width and crawl stroke times. These findings may all be related to the fact that swimmers with broad shoulders are better suited for high power output in the water (Mazza *et al.*, 1994). Together with this it can also be expected that swimmers will have proportionally higher muscle mass values when compared to the Phantom. High muscular profiles among swimmers will probably tend to benefit power and explosive power. This notion is supported by Heyward and Stolarczyk (1996) who stated that high muscle mass percentages will contribute towards an increase in power.

Research has provided a possible explanation for the result that female swimmers displayed a proportionally larger average thigh girth than the Phantom. Gross *et al.* (2000) concluded that thigh girth was among other variables significantly related to females' maximum lifting capacity. The assumption that can, therefore, be made is that swimmers would display larger thigh girths due to the fact that they have to generate a high level of leg propulsion in the water and would need a higher level of leg strength to do that. In one study on male and female swimmers, Smith (1978) also showed that females depended more on their kicking speed for forward propulsion than the males who depended more on their arm pulling speed. The female swimmers did, however, also show higher fat mass and percentage values than the Phantom. It is common knowledge that an increase in body fat will improve buoyancy (Maglischo, 2003), which will in turn lead to less drag experienced by the swimmer in the water, improved efficiency of the swimmer and a lower metabolic cost of swimming (Chatard *et al.*, 1990, 1995;

McLean and Hinrichs, 2000). However, Maglischo (2003) points out that more body fat may also possibly result in an increased body surface area, which may lead to increased body shape drag forces and decrease swimming times.

Moreover, the data clearly indicate that male swimmers had proportionally larger neck, flexed arm and forearm girths than the Phantom. Mayhew *et al.* (1993) found high relationships between trained athletes' arm circumferences and muscle cross-sectional areas and one-repetition maximum (1-RM) lifts in the bench press, squat and dead lift. It is possible that the higher larger girth values are again related to swimmers' better developed strength capacity and ability to generate swimming propulsion through their upper arms and trunks. This might also explain the proportionally larger transverse chest, chest depth, humerus and bideltiod breadth, longer tibial height and larger lean body mass values that were observed in the swimmers' group when compared to the Phantom.

With regard to the existence of sexual dimorphism among the male and female swimmers, the separation in standard error lines of Figures 17.1–17.5 shows that only 13 variables (out of a possible 44) were proportional significantly bigger in one gender compared to the other. Males were proportionally bigger compared to females in the following variables: neck and wrist girths as well as humerus breadth. As expected, the females were proportional significantly bigger in triceps, supraspinale, front thigh and medial calf skinfolds; head, gluteal and thigh girths; bitrochanteric breadth as well as fat mass and percentage compared to the males. These findings support those of related research conducted by De Ridder *et al.* (2003), who also demonstrated that significant differences between the genders with regard to their anthropometric measurements become insignificant during the proportional comparisons of athletes. The similar pattern of the Z-values for most of the variables between the two genders also indicated that a great deal of homogeneity exists with regard to the anthropometric variables of swimmers. This was especially evident for the Z-values of the non-significant different skinfold, length and breadth variables.

The majority of variables that did not display a lot of homogeneity between the two gender groups were the body composition variables, especially the skinfold measurements, fat mass and percentage. These results are in accordance with the findings of McLean and Hinrichs (2000), who also found that female swimmers display a much higher fat mass and higher skinfold measurements than the males. This finding may also help to explain the proportional significantly bigger gluteal and thigh girths of the females. Wilmore and Costill (2004) concluded that fat deposition of females increases in the thigh and hip areas due to increased lipoprotein lipase activity in these areas. A higher fat deposition in these areas will, therefore, lead to a proportional increase in the gluteal and thigh girths of females compared to males who carry more fat in the abdomen and upper body (Wilmore and Costill, 2004).

It is difficult to explain from the literature why significantly bigger proportional average neck girths were displayed by the male compared to the female swimmers. It may be that males spend more time in resistance training than females. Males in this study indicated that they usually spent about 3 hours/week in resistance training compared to the females who generally spend 1.6 hours/week in resistance training. Muscles such as the trapezius and sternocleidomastoid which lie in the neck area and have an effect on the neck girth may well develop more due to the mentioned training time differences in males compared to females. It is, however, difficult to explain why the rest of the proportional muscle related girth areas did not show the same trend with regard to gender differences as the neck girth.

According to the literature, differences in the bone growth patterns between the two genders could possibly explain the significant differences between the proportional wrist girth and humerus breadth variables in the two populations (Malina *et al.*, 2004). The last-mentioned authors indicated that boys generally reach higher cortical bone widths during late adolescence compared to girls. Concurring with the bitrochanteric findings of the present study, Moore and Dalley (1999) found that the female pelvis was wider, shallower and the hip bones further apart compared to the male pelvis which was narrower, deeper and the hip bones nearer to each other. The finding that the females obtained a significantly bigger average proportional head girth compared to the males was surprising in view of previous research findings that did not coincide with these results. Mazza *et al.* (1994) had, for example, concluded that male crawl-stroke swimmers generally show bigger overall girth values than female crawl-stroke swimmers.

The above-mentioned results seem to support the observation of Gualdi-Russo and Graziani (1993) that the greatest homogeneity exists between genders who participate in sport where the same activities are performed on a daily basis and in competitions. The study does, however, show that sexual dimorphism does exist for certain anthropometric variables between female and male swimmers. These anthropometric differences could partially explain the differences in arm coordination and swimming performances between the two genders (McLean and Hinrichs, 2000; Seifert *et al.*, 2004).

5 CONCLUSION

In conclusion, significant differences were found between the South African female and male crawl stroke swimmers in almost all of the absolute anthropometric variables (44 out of a possible 51 variables) except for the pectoral and abdominal skinfolds; gluteal and thigh girths; biiliocristal and bitrochanteric breadths as well as ectomorphy when comparisons were made. These differences became insignificant when comparisons were made in a proportional manner between the two genders. Only 13 variables (out of a possible 44) were proportionally significantly bigger in one gender compared to the other. Significantly bigger values were seen for the neck girth and humerus breadth when the males were compared to the females, who recorded significantly bigger triceps, supraspinale, front thigh and medial calf skinfolds; head, gluteal and thigh girths; bitrochanteric breadths as well as fat mass and percentages compared to the males.

The anthropometric proportionality comparison between the swimmers of both genders and the Phantom showed that participants in the crawl stroke swimming event were very similar in appearance. Compared to the Phantom the swimmers were collectively proportionally larger in chest girth, longer in armspan, arm, thigh and tibia length, larger in biacromial breadth as well as in the muscle mass value. The swimmers did, however, display smaller values (negative) in all the other skinfold, girth, length, height, breadth and body composition variables when they were proportionally compared to the Phantom. With regards to gender-group the female swimmers were larger in thigh girth, fat mass and fat mass percentage, while the male swimmers were proportional larger in flexed arm, and waist girth, larger in transverse chest, chest depth, humerus and bideltiod breadth, longer in tibial height, larger in lean body mass and similar in neck and forearm girth.

It can, therefore, be concluded that sexual dimorphism becomes less obvious when comparing the male and female crawl-stroke swimmers proportionally. The patterns of each of the anthropometric proportionality profiles of the female and male swimmers are very similar when they are compared to the Phantom scores. These patterns indicate that at elite levels of participation, both male and female swimmers develop a typical and unique morphology for their particular sport, in this instance crawl-stroke swimming. Despite this, the female swimmers still appear to be proportionally less robust in the upper body and appear to have a proportionally greater adipose tissue deposition on the limbs and waist compared to the male swimmers. These anthropometric differences that do, therefore, still exist after proportional comparisons between the two gender groups may still have certain implications for the kinematic and performance parameter differences that are seen. In spite then of the attempt to proportionally quantify differences in anthropometric characteristics between female and male crawl-stroke swimmers, female swimmers still exhibit certain anthropometric traits that are specific to them. These gender-specific anthropometric characteristics may be related to the specific responses of different hormones to the swim training regiments that are followed. As been mentioned before, females for example gain substantially less muscle mass in response to a given training stimulus than males do, probably because of lower levels of testosterone. Furthermore, research has also shown that swim training seems to act in different ways on the serum lipid and lipoprotein levels of female and male competitive swimmers, which also give rise to gender-specific subcutaneous fat distributions (Folin *et al.*, 1991).

Several shortcomings of this study should, however, be considered when interpreting the data. The small group sizes in this study could have caused outliers to have influenced the mean values of the respective anthropometric measurements more than would have been the case with larger group sizes. Generalization of the results to the whole of South Africa may, therefore, not be accurate. In view of the fact that some of the measures for calculation of the Phantom scores were derived from old data (1969, 1970, 1971 and 1974) and data of people in other parts of the world than South Africa (Ross and Marfell-Jones, 1991), it can be recommended that new data of the population in South Africa should rather be used to calculate the Phantom scores. This will ensure that the proportional anthropometric quantification of South African swimmers are more accurate and relevant to the current, general population of South Africa.

6 REFERENCES

Ackland, T.R. and Mazza, J.C., 1994, Introduction. In *Kinanthropometry in Aquatic Sports*, edited by Carter, J.E.L., Ackland, T.R., Mazza, J.C. and Ross, W.D. (Champaign, Ill.: Human Kinetics Publishers), pp. viii–x.

Åstrand, P., Rodahl, K., Dahl, H.A. and Stromme, S.B., 2003, *Textbook of Work Physiology: Physiological Bases of Exercise*, 4th ed (Champaign, Ill.: Human Kinetics Publishers).

Bell, R.D. and Hoshizaki, T.B., 1981, Relationships of age and sex with range of motion of seventeen joint actions in humans. *Canadian Journal of Applied Sport Sciences*, **6**, 202–206.

Bishop, P., Cureton, K. and Collins, M., 1987, Sex difference in muscular strength in equally-trained men and women. *Ergonomics*, **30**, 675–687.

Carter, J.E.L., Ross, W.D., Aubry, S.P., Hebbelinck, M. and Borms, J., 1982, Anthropometry of Montreal Olympic athletes. In *Physical Structure of Olympic Athletes, Part I: The Montreal Olympic Games Anthropological Project*, edited by Carter, J.E.L. (Basel: S Kerger Press), pp. 25–52.

Carter, J.E.L. 1984, Somatotypes of Olympic athletes from 1948 to 1976. In *Physical Structure of Olympic Athletes, Part II. Kinanthropometry of Olympic athletes*, edited by Carter, J.E.L. (Basel: S Kerger Press), pp. 80–109.

Carter, J.E.L. and Heath, B.H., 1990, *Somatotyping – Development and Applications* (Cambridge, NY.: Cambridge University Press).

Carter, J.E.L. and Marfell-Jones, M.J., 1994, Somatotypes. In *Kinanthropometry in Aquatic Sports*, edited by Carter, J.E.L., Ackland, T.R., Mazza, J.C. and Ross, W.D. (Champaign, Ill.: Human Kinetics Publishers), pp. 55–82.

Carter, J.E.L. and Ackland, T.R., 1998, Sexual dimorphism in the physiques of World Championship divers. *Journal of Sports Sciences*, **16**, 317–329.

Chatard, J.C., Lavoie, J.M. and Lacour, J.R., 1990, Analysis of determinants of swimming economy in front crawl. *European Journal of Applied Physiology and Occupational Physiology*, **61**, 88–92.

Chatard, J.C., Senegas, X., Selles, M., Dreanot, P. and Geyssant, A., 1995, Wet suit effect: a comparison between competitive swimmers and triathletes. *Medicine and Science in Sports and Exercise*, **27**, 580–586.

Coetzee, B., 2002, Somatotiperingsprofiele van manlike en vroulike Suid-Afrikaanse kruipslagswemmers (Somatotype profiles of male and female South African crawl stroke swimmers). *South African Journal for Research in Sport, Physical Education and Recreation*, 24, 1–12.

De Ridder, J.H., Smith, E., Wilders, C. and Underhay, C., 2003, Sexual dimorphism in elite middle-distance runners: 1995 All-Africa Games (Project HAAGKIP). *Kinanthropometry 7: Proceedings of the Seventh Scientific Conference of the International Society for the Advancement of Kinanthropometry*, Brisbane, 7–12 September, 2000, edited by De Ridder, J.H. and Olds, T. (Potchefstroom, South Africa: Potchefstroom University for Christian Higher Education), pp. 57–75.

Drinkwater, D.T. and Mazza, J.C., 1994, Body composition. In *Kinanthropometry in Aquatic Sports*, edited by Carter, J.E.L., Ackland, T.R., Mazza, J.C. and Ross, W.D. (Champaign, Ill.: Human Kinetics Publishers), pp. 102–137.

Folin, M., Contiero, E., Marin, V. and Parnigotto, P.P., 1991, Serum lipoprotein levels and body composition in a group of competitive swimmers. *Medicina Dello Sport*, **44**, 233–241.

Gatford, K.L., Egan, A.R., Clarke, I.J. and Owens, P.C., 1998, Sexual dimorphism of the somatotrophic axis. *The Journal of Endocrinology*, **157**, 373–389.

Geladas, N.D., Nassis, G.P. and Pavlicevic, S., 2005, Somatic and physical traits affecting sprint swimming performance in young swimmers. *International Journal of Sports Medicine*, **26**, 139–144.

Grimston, S.K. and Hay, J.G., 1986, Relationship among anthropometric and stroking characteristics of college swimmers. *Medicine and Science in Sports and Exercise*, **18**, 60–68.

Gross, M.T., Dailey, E.S., Dalton, M.D., Lee, A.K., McKiernan, T.L., Vernon, W.L. and Walden, A.C., 2000, Relationship between lifting capacity and anthropometric measures. *The Journal of Orthopaedic and Sports Physical Therapy*, **30**, 237–247.

Gualdi-Russo, E. and Graziani, I., 1993, Anthropometric somatotype of Italian sport participants. *Journal of Sports Medicine and Physical Fitness*, **33**, 282–291.

Heyward, V.H. and Stolarczyk, L.M., 1996, *Applied Body Composition Assessment*. (Champaign, Ill.: Human Kinetics Publishers).

Hollander, A.P., De Groot, G., Van Ingen Schenau, G.J., Kahman, R. and Toussaint, H.M., 1988, Contribution of the legs to propulsion in front crawl swimming. In *Proceedings of the 5th International Symposium of Biomechanics and Medicine in Swimming, Swimming Science V* (Champaign, Ill.: Human Kinetics Publishers), pp. 39–43.

Jiang, J., 1993, How to select potential olympic swimmers. *American Swimming Magazine*, pp. 14–18.

Kilbride, E., McLoughlin, P., Gallagher, C.G. and Harty, H.R., 2003, Do gender differences exist in the ventilatory response to progressive exercise in males and females of average fitness? *European Journal of Applied Physiology*, **89**, 595–602.

Kokkalis, C.Z., Athanassaki, M., Gourgoulianis, K.I. and Molyvdas, P.A., 2001, Diet, lung function and swimmers' performance. *Journal of Nutritional and Environmental Medicine*, **11**, 121–125.

Lafortuna, C.L. and Passerini, L., 1995, A new instrument for the measurement of rib cage and abdomen circumference variation in respiration at rest and during exercise. *European Journal of Applied Physiology and Occupational Physiology*, **71**, 259–65.

Maglischo, E.W., 2003, *Swimming fastest* (Champaign, Ill.: Human Kinetics Publishers).

Malina, R.M., Bouchard, C. and Bar-Or, O., 2004, *Growth, maturation and Physical Activity*, 2nd ed. (Champaign, Ill.: Human Kinetics Publishers).

Martin, A.D., Spenst, L.F., Drinkwater, D.T. and Clarys, J.P., 1990, Anthropometric estimation of muscle mass in men. *Medicine and Science in Sports and Exercise*, **22**, 729–733.

Mazza, J.C., Ackland, T.R., Bach, T.M. and Cosolito, P., 1994, Absolute body size. In *Kinanthropometry in Aquatic Sports*, edited by Carter, J.E.L., Ackland, T.R., Mazza, J.C. and Ross, W.D. (Champaign, Ill.: Human Kinetics Publishers), pp. 15–54.

Mayhew, J.L., Piper, F.C. and Ware, J.S., 1993, Anthropometric correlates with strength performance among resistance trained athletes. *Journal of Sports Medicine and Physical Fitness*, **33**, 159–165.

McLean, S.P. and Hinrichs, R.N., 2000, Buoyancy, gender, and swimming performance. *Journal of Applied Biomechanics*, **16**, 248–263.

Montpetit, R.M. and Smith, H., 1988, Build for speed. *Swimming Technique*, **24**, 30–32.

Moore, K.L. and Dalley II, A.F., 1999, *Clinically Oriented Anatomy*, 4th ed., (Philadelphia: Lippincott Williams and Wilkins).

Nindl, B.C., Scoville, C.R., Sheehan, K.M., Leone, C.D. and Mello, R.P., 2002, Gender differences in regional body composition and somatotrophic influences of IGF-I and leptin. *Journal of Applied Physiology*, **92**, 1611–1618.

Norton, K.I., Whittingham, N., Carter, L., Kerr, D., Gore, G. and Marfell-Jones, M., 1996a, Measurement techniques in anthropometry. In *Anthropometrica: A Textbook of Body Measurements for Sports and Health Courses*, edited by Norton, K.L. and Olds, T.S. (Marrickville, NSW: Southwood Press), pp. 25–73.

Norton, K.I., Olds, T.S., Olive, S.C. and Craig, N.P., 1996b, Anthropometry and sports performance. In *Anthropometrica*, edited by Norton, K.L. and Olds, T.S. (Marrickville, NSW: Southwood Press), pp. 287–364.

Pelayo, P., Sidney, M., Kherif, T., Chollet, D. and Tourny, C., 1996, Stroking characteristics in freestyle swimming and relationships with anthropometric characteristics. *Journal of Applied Biomechanics*, **12**, 197–205.

Pinkstaff, C.A., 1998, Salivary gland sexual dimorphism: a brief review. *European Journal of Morphology*, **36**, 31–34.

Rosenbaum, M. and Leibel, R.L., 1999, Clinical review 107: Role of gonadal steroids in the sexual dimorphisms in body composition and circulating concentrations of leptin. *The Journal of Clinical Endocrinology and Metabolism*, **84**, 1784–1789.

Ross, W.D., Leahy, R.M., Drinkwater, D.T. and Swenson, P.L., 1981, Proportionality and body composition in male and female Olympic athletes: a kinanthropometric overview. In *Female Athlete: A Socio-Psychological and Kinanthropometric Approach*, edited by Borms, J., Hebbelinck, M. and Venerando, A. (Basel: S. Karger), pp. 74–84.

Ross, W.D., Ward, R., Leahy, R.M. and Day, J.A.P., 1982, Proportionality of Montreal athletes. In *Physical Structure of Olympic Athletes, Part I: The Montreal Olympic Games Anthropological Project*, edited by Carter, J.E.L. (Basel: S Kerger Press), pp. 81–106.

Ross, W.D. and Ward, R., 1984, Proportionality of Olympic athletes. In *Physical Structure of Olympic Athletes, Part II: Kinanthropometry of Olympic athletes*, edited by Carter, J.E.L. (Basel: S Kerger Press), pp. 110–143.

Ross, W.D. and Marfell-Jones, M.J., 1991, Kinanthropometry. In *Physiological Testing of High-Performance Athletes*, edited by MacDougall, J.D., Wenger, H.A. and Green, H.J. (Champaign, Ill.: Human Kinetics Publishers), pp. 223–308.

Ross, W.D., Leahy, R.M., Mazza, J.C. and Drinkwater, D.T., 1994, Relative body size. In *Kinanthropometry in Aquatic Sports*, edited by Carter, J.E.L., Ackland, T.R., Mazza, J.C. and Ross, W.D. (Champaign, Ill.: Human Kinetics Publishers), pp. 83–101.

Seifert, L., Boulesteix, L. and Chollet, D., 2004, Effect of gender on the adaptation of arm coordination in front crawl. *International Journal of Sports Medicine*, **25**, 217–223.

Shephard, R.J., 2000, Exercise and training in women, Part I: Influence of gender on exercise and training responses. *Canadian Journal of Applied Physiology*, **25**, 19–34.

Siders, W.A., Bolonchuk, W.W. and Lukaski, H.C., 1991, Effects of participation in a collegiate sport season on body composition. *Journal of Sports Medicine and Physical Fitness*, **31**, 571–576.

Simmons, S.E.C., 2003, *Sexual Dimorphism and the Correlates of Sprint Swim Performance* (University of Oregon: Kinesiology Publications).

Smith, L.E., 1978, Anthropometric measurements, and arm and leg speed performance of male and female swimmers as predictors of swim speed. *Journal of Sports Medicine and Physical Fitness*, **18**, 153–168.

StatSoft, Inc., 2005, STATISTICA (data analysis software system), version 7, web: www.statsoft.com.

Takagi, H., Sugimoto, S., Nishijima, N. and Wilson, B., 2004, Differences in stroke phases, arm-leg coordination and velocity fluctuation due to event, gender and performance level in breaststroke. *Sports Biomechanics*, **3**, 15–27.

Tyndall, G.L., Kobe, R.W. and Houmard, J.A., 1996, Cortisol, testosterone, and insulin action during intense swimming training in humans. *European Journal of Applied Physiology and Occupational Physiology*, **73**, 61–65.

Vervaecke, H. and Persyn, U., 1981, Some differences between men and woman in various factors which determine swimming performance. *In The Female Athlete*, edited by Borms, J., Hebbelinck, M. and Venerando, A. (Basel: S Kerger Press), pp. 150–156.

Watkins, J. and Gordon, A.T., 1983, The effects of leg action on performance in the sprint front crawl stroke. In *Biomechanics and Medicine in Swimming*, edited by Hollander, A.P., Huijing, P.A. and De Groot, G. (Champaign, Ill.: Human Kinetics Publishers), pp. 310–314.

Wilmore, J.H. and Costill, D.L., 2004, *Physiology of Sport and Exercise*, 3rd ed. (Champaign, Ill.: Human Kinetics Publishers).

Withers, R.T., LaForgia, J., Pillans, R.K., Shipp, N.J., Chatterton, B.E., Schultz, C.G. and Leaney, F., 1998, Comparisons of two-, three-, and four-compartment models of body composition analysis in men and women. *Journal of Applied Physiology*, **85**, 238–245.

Withers, R.T., Whittingham, N.O., Norton, K.I., La Forgia, J., Ellis, M.W. and Crockett, A., 1987a, Relative body fat and anthropometric prediction of body density of female athletes. *European Journal of Applied Physiology and Occupational Physiology*, **56**, 169–180.

Withers, R.T., Whittingham, N.O., Norton, K.I., La Forgia, J., Ellis, M.W. and Crockett, A., 1987b, Relative body fat and anthropometric prediction of body density of male athletes. *European Journal of Applied Physiology and Occupational Physiology*, **56**, 191–200.

Index